EARTHLY GOO

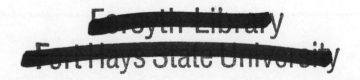

EARTHLY GOODS

Environmental Change and Social Justice

EDITED BY

FEN OSLER HAMPSON
and JUDITH REPPY

Cornell University Press

ITHACA AND LONDON

First published 1996 by Cornell University Press.

Library of Congress Cataloging-in-Publication Data

Earthly goods : environmental change and social justice / edited by
   Fen Osler Hampson, Judith Reppy.
      p.   cm.
   Includes bibliographical references and index.
   ISBN 0-8014-3289-8 (cloth : alk. paper).—ISBN 0-8014-8362-X
(paper : alk. paper)
      1. Environmental policy—Economic aspects.   2. Social justice.
   3. Conservation of natural resources.   I. Hampson, Fen Osler.
   II. Reppy, Judith, 1937–   .
   HC79.E5E17   1996
   333.7—dc20   96-19784

Printed in the United States of America

This book is printed on Lyons Falls Turin Book,
a paper that is totally chlorine-free and acid-free.

Contents

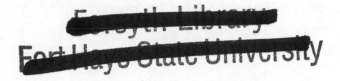

Contributors   vii

Preface   ix

Introduction: Framing the Debate   1
Fen Osler Hampson, Pierre Laberge,
and Judith Reppy

1 Environmental Change and the Varieties of Justice   9
Henry Shue

2 Concepts of Community and Social Justice   30
Will Kymlicka

3 Inherent Value and Moral Standing in
Environmental Change   52
Wendy Donner

4 Societies in Space and Place   75
Iain Wallace and David B. Knight

5   The Normative Structure of International Society   96
        Christian Reus-Smit

6   Impoverishment and the National State   122
        Joseph Camilleri

7   Social Movements, Ecology, and Justice   154
        Smitu Kothari

8   Science and Norms in Global Environmental
    Regimes   173
        Sheila Jasanoff

9   Campaigning and Critique: Public-Interest Groups
    and Environmental Change   198
        Steven Yearley

10 Breathing Room: Negotiations on Climate Change   221
        Peter Timmerman

    Conclusion: Liberalism Is Not Enough   245
        Judith Reppy and Fen Osler Hampson
    Index   257

Contributors

JOSEPH CAMILLERI is professor of politics in the School of Social Sciences at LaTrobe University, Bundoora.

WENDY DONNER is professor of philosophy at Carleton University in Ottawa.

FEN OSLER HAMPSON is professor of politics at the Norman Paterson School of International Affairs at Carleton University.

SHEILA JASANOFF is professor and chair of the Department of Science and Technology Studies at Cornell University, Ithaca, N.Y.

DAVID B. KNIGHT is dean of the College of Social Science at the University of Guelph, Ontario.

SMITU KOTHARI is director and editor of *Lokayan* in Delhi.

WILL KYMLICKA is professor of philosophy at the University of Ottawa.

PIERRE LABERGE is professor of philosophy at the University of Ottawa.

JUDITH REPPY is associate professor of science and technology studies and associate director of the Peace Studies Program at Cornell University.

CHRISTIAN REUS-SMIT is assistant professor of politics at Monash University, Clayton.

HENRY SHUE is the Wyn and William Y. Huchinson Professor and director of Ethics and Public Life at Cornell University.

PETER TIMMERMAN is a research associate at the International Federation of Institutes for Advanced Study/Institute for Environmental Studies at the University of Toronto.

IAIN WALLACE is professor of geography at Carleton University.

STEVEN YEARLEY is professor and chair of the Sociology Department at York University, England.

Preface

THIS book has its origins in a 1992–93 series of workshops and a conference on global environmental change and social justice organized jointly by the Peace Studies Program at Cornell University, Carleton University, and Université d'Ottawa. Our aim was to bring together scholars from a range of disciplines to tackle together the knotty issues raised for environmental policymaking when social justice concerns are taken seriously. Participants included philosophers, political scientists, sociologists, historians, geographers, economists, biological scientists, and lawyers. Construction of a shared discourse among so many disciplines did not come easily, but we were rewarded in the final conference by a sustained discussion of the issues raised in this book. Most of the papers included here were presented at both a workshop and the final conference, and all have been revised in response to group discussion.

The book covers four major themes. Henry Shue, Wendy Donner, Will Kymlicka, and Iain Wallace and David Knight lay out alternative frameworks for evaluating social justice from different philosophical perspectives. Christian Reus-Smit, Joseph Camilleri, and Smitu Kothari discuss the role of the state and of substate actors in the international politics of the environment. Sheila Jasanoff and Steven Yearley take up the question of the role of science in framing the debate on global environmental change and the use of science as a resource by various actors in actual negotiations. Finally, Peter Timmerman provides an account of international negotiations in which the themes of

the previous chapters are developed and used to argue for the centrality of social justice in reaching desirable outcomes.

The project was supported by funds from grants to Cornell's Peace
Studies Program from the John D. and Catherine T. MacArthur Foundation and the National Science Foundation; a grant from the Social
Sciences and Humanities Research Council of Canada; and additional
support from the Faculty of Arts and Graduate Studies and Research,
Carleton University, Université d'Ottawa, and the Social Science Research Council. It was in every respect a collaborative effort, and we
thank our colleagues in both the United States and Canada for their
many helpful suggestions and hard work. We also thank Janice Fo-
chuk of Carleton University and Elaine Scott and Sandra Kisner of the
Peace Studies Program at Cornell University for providing skilled assistance in organizing the workshops and conferences and coordinating this multinational project.

F. O. H.
J. R.

EARTHLY GOODS

Introduction:
Framing the Debate

FEN OSLER HAMPSON, PIERRE LABERGE,
and JUDITH REPPY

THE specter of global warming haunts the Earth's future. We are already familiar with the problem of ozone depletion in the upper atmosphere caused by the release of manmade chemicals. In the next century the world will experience climate changes on a global scale because of the buildup of "greenhouse gases" in the atmosphere from the burning of fossil fuels and the destruction of the Earth's forests. Although the exact timing and magnitude of the changes cannot be predicted, they are likely to leave no country untouched. At the same time deforestation and other human activities are leading to the extinction of many plant and animal species and a concomitant reduction in the planet's genetic stockpile. What principles should guide our response to these undesirable consequences of human activity? We argue that a concern for social justice is central to devising acceptable policies for a global response to environmental change and should be placed at the forefront of the international debate.

To a large extent this concern for social justice is already present. Although warnings from the scientific community and environmental groups were responsible for catapulting global warming to the forefront of the international agenda, international negotiations have not centered on scientific details so much as on who should do what to halt and reverse the ecological damage that imperils the future quality of life. In other words, the negotiations, such as the 1992 Rio Conference on the Environment and Sustainable Development, have become an important arena for competing claims and counterclaims about who should assume moral responsibility for cleaning up the environ-

I

ment and preventing future harm; scientific debate on the extent of global warming and its impacts may never reach closure, but the political debate over the distribution of the costs is sure to persist.

HARD CHOICES

The issues of social justice that arise around global environmental change are far more complex than is generally conceded in ordinary political discourse. For example, scientists tell us that the global total of greenhouse-gas emissions must be reduced below current total levels if serious global warming is to be avoided—by some estimates up to 60 percent of current emissions. What is a fair allocation of the costs of preventing global warming? Should emissions be divided up on a per capita basis—which would favor countries with expanding populations—in proportion to present emission levels, in proportion to expected future emission levels, according to levels of GNP, or some other formula?

The answer to this question is not narrowly economic or technical. If we ask all countries to reduce emissions by the same percentage, we may condemn Third World countries to perpetual poverty because they would have to forgo the use of hydrocarbons in their economic development. Even though alternative energy substitutes might be available, at current energy prices they are not competitive. Alternatively, if we adopt a per capita measure for allocating emission requirements, rich countries would face enormous sacrifices because they would bear the brunt of reductions. For example, if scientists tell us that we need to reduce overall emissions by 20 percent, then rich countries will have to reduce their emission by much more than 20 percent in order for poorer countries to raise energy consumption to meet their growth requirements. An allocation of allowable emissions according to a "polluter pays" principle again poses a moral dilemma. Should current generations bear moral responsibility for actions taken by their forebears? If not them, what about future generations? Should our children and grandchildren be forced to live in a world marred by the consequences of global warming?

If we knew the physical and biological consequences of global warming were negligible or modest, the answer would be easy. Similarly, if the costs of disruption were enormous, the moral imperative to take preventive actions now would be great. The range of uncertainty about the effects of global warming on ecological systems is considerable, however, and the range of uncertainty about its possible economic and social consequences and the ability of institutions to adapt to environmental change is even greater. Thus the real question is what to do when confronted with a problem that has so many uncertainties associated with it. Whatever choice one makes, a "business-as-usual" ethic is clearly unacceptable in dealing with a problem that has potentially global costs associated with it.

In another example, indigenous peoples in the Chittagong Hill Tracts of Bangladesh are being overrun by settlers who have been encouraged by their government to settle and develop less populated, frontier homelands. The same thing has happened to indigenous peoples around the globe from Amazonia to Indonesia. Whose rights should be respected? Those of indigenous inhabitants who claim a right to control their traditional homelands or those of the nonindigenous poor who are landless, starving, and have nowhere else to go?

Most people are inclined to side with the moral claims of indigenous peoples, who are the most obvious victims of economic progress and development. Their prior claim—that "they were there first"—has intrinsic moral appeal. So too does the fact that their culture and way of life rests on their relationship with their natural surroundings. But on the other hand, the new settlers are also victims of development processes and circumstance. In the Chittagong case, the government of Bangladesh resettled those people driven from the coastal areas by floods in the hill tracts. In a victim-against-victim dispute, whose moral claims are more valid? Those of the settler or those of the indigenous person? The principle of individual equality, that is, the view that all have equal moral standing, does not allow us to assign special value to the indigenous way of life. Should we? Is the culture of indigenous peoples worth preserving? Do they have special moral claims because of their "ecologically correct" way of living?

Environmental activists and some native groups fought the decision of the British Columbia government to allow clear-cut logging in some parts of Clayoquot Sound on Vancouver Island—one of the last remaining stands of old-growth rain forest in the world. Their view is that any development of Clayoquot will lead to the destruction of this natural habitat. On the other hand, "doing right" by the forest will jeopardize the survival of the local community, which depends on logging. We may also wish to assert a moral claim on behalf of the forest ecosystem. Whose interests should prevail: those of the local community or those of environmentalists motivated by a biospheric ethic opposed to development?

This example presents a different moral dilemma. Does the environment have an intrinsic moral claim that supersedes the interests of humans who depend on the exploitation of the environment for their livelihood? Are old-growth forests worth keeping, given that replanting and sustainable cropping is a real possibility? More broadly, the issue of moral standing for other species must be weighed against a human-centered ethic. To date, a standard that accepts the need to preserve certain species has been applied intermittently. For every species that has gained protection, there are dozens of others that have not. Which species are worthy of protection? In the case of the forest, the argument can be made that a whole ecosystem deserves protection. One can also argue, as some have, for the need to protect old-growth forests for the pleasure and enjoyment of future generations. The latter argument, however, relies on human-centered ethical standards, not biological or ecocentric ones. The difficulty in finding a consistent line of argument is apparent. A human-centered ethic that assigned value to the pleasure given by old-growth forest would have to reconcile itself with the moral claims of the loggers who are alive today and whose welfare (and their children's) depends upon cutting the forest.

A final example concerns population growth, which lies at the heart of many of the problems of environmental degradation and resource depletion around the world. Whole societies find it difficult to break out of a cycle of poverty in which spiraling population growth overwhelms their capacity to improve health, education, and employment opportunities, with resultant pressure on environmental resources. Policies to manage population growth, however, have too often focused crudely on controlling women's choices, without recognizing the principle that women should have a voice in decisions that affect them. Is there a point at which the danger of population growth to the welfare of the society overrides individual women's preferences? If so, how can environmental concerns be reconciled with the moral imperative to value human life?

The challenge here is to find solutions that respect individual rights to make reproductive choices while encouraging voluntary limitation of family size. But women typically lack access to and standing in the power structures that determine the conditions of their lives. Moreover, it is clear that in many societies women are responsible for the activities that are altering the environment, whether it be cooking on charcoal braziers in the favelas of cities plagued by deadly levels of air pollution or tending the rice paddies of Southeast Asia. Thus whether the question is population management or effective policy to reduce environmentally damaging practices, the cooperation and participation of women is necessary. But it may not be the case that including women in the planning and implementation of development policy will necessarily lead to ecologically sound policies. What should our choice be if grassroots approaches that value broad participation and equality for women produce policies and practices that increase population growth rates or expand environmental degradation?

These examples illustrate the point that virtually all environmental decisions raise ethical dilemmas. The issues of whose preferences should receive attention or of whose rights should have priority in devising "just" solutions to pressing environmental concerns are questions not easily addressed. What does "fairness" mean in dividing responsibilities between rich and poor nations regarding problems like global warming, whose causes are uncertain and whose consequences may not be experienced for decades to come? How should societal procedures for addressing these ethical issues be structured?

At a deeper level the ethical issues rest on our understanding of humanity's place in nature. Our judgment of good and bad, just and unjust, needs to be informed by a sense of our connection to the natural world. To the degree that we value the ecosystem for its own sake and not simply as an exploitable resource, we enter a new realm of ethical thinking, one in which justice is not defined only in terms of an equitable distribution of costs and benefits among humans, but also in terms of the effects of human activities on the whole ecosystem. That is, we are no longer able to treat the question of justice as a problem in specifying a calculus of trade-offs separable from the environment itself; some elements of the environment may deserve to be protected absolutely because of their importance in sustaining the ecosystem.

It has become clear that concepts of social justice rooted in traditional liberal theory are not adequate for addressing the kinds of competing moral claims that have arisen over environmental issues. The interests of disenfranchised majorities—like the world's poor who live at or below the subsistence level, or those of unborn generations—are problematic from the point of view of liberal theories, which start with the assumption that individual preferences are equal and should simply be taken at face value. This stance is too passive when it comes to the environment, not only because some parties are not represented at the bargaining table, but because the preferences of those who are represented are often at the root of the problem. Current energy consumption patterns, for example, are key factors in the accumulation of greenhouse gases that contribute to global warming. A just solution to the problem must not only seek to limit future emissions through an equitable formula, but also to distribute the costs of preventive or adaptive solutions according to agreed-on standards of fairness—for example, by assigning the costs to those who have contributed most to the problem in the past.

The role of community and culture is also not easily captured in traditional liberal theories of social justice, even in those that accept that all preferences may not be equal, because the theories, which are typically grounded in a concern for individual rights and preferences, do not recognize claims based on the value of group identity. To accommodate the claims of indigenous cultures, for example, we need a theory that gives standing to communities and other collectivities.

ALTERNATIVE PERSPECTIVES

The debate about justice in the context of global environmental change, however, is not just about moral standards but about the fundamental nature and scope of justice. At one level, the debate centers on the rights of human beings versus those of other species who inhabit the planet. Some theorists, for instance, argue that the liberal definition of the moral community is too narrow and must be expanded to take into account the interests of animals and plants. These theorists offer a vision of the moral community that goes beyond human-centered ethics to an extensionist or even to a holistic value paradigm centered on the ecosystem. According to this view, human beings have obligations not just to themselves but to nature. When the existence of other species is threatened by human activity, human interests must defer to those of endangered species.

For those who work within a more traditional, human-centered view of justice, the question is not so much one of how to go about assigning value to sentient beings or even to nonsentient things, but how to evaluate preferences and incorporate the interests of different communities into the concept of social justice. One of the striking aspects of the debate surrounding global environmental change is that new concepts of space, community, and culture are informing discussions about social justice. There is growing recognition that geographically defined communities in the hinterland have special moral significance because of their close dependence on environmental resources for their livelihood; this dependence, which is usually cloaked in modern industrial society, is transparent in the lives of these communities. Their claims on the environment are sometimes intimately related to their survival as cultural and even national entities. Nowhere is this problem more acute than in the case of indigenous peoples. These groups find themselves in a highly vulnerable and disadvantageous position. Their very existence is threatened. And prevailing notions of social justice cannot accommodate them.

Just as new concepts of space, community, and culture inform the debate about environmental change and social justice, traditional notions about the moral purpose of the state are also coming under scrutiny. Is it even possible to construct just distributive arrangements to tackle the root causes of environmental decline if the modern society of states reponds only to claims cast in terms of economic progress? According to some scholars, the roots of our environmental crisis lie in modern industrial capitalism with its emphasis on economic growth, an emphasis that informs the political context in which environmental issues are negotiated, privileging some actors and arguments while excluding others. They argue that without new social structures and participatory forms of governance, a genuine moral discourse directed at changing and influencing the underlying norms of current behavior will not be possible. The task of achieving just and fair solutions to the current environmental crisis requires the participation of voices that are now effectively silenced.

A similar argument can be made about the scientific discourse that has been so influential in framing the debate over global environmental change. At one level, scientific discourse translates into a kind of market discourse because it denies the existence or importance of the national state and other identities or group affiliations by its claims to universality and objectivity. At another level, science is crucial to the state because it has become the adopted discourse of the most powerful and wealthy states and forms the basis and legitimacy for key policy decisions. Science's claim to universality has particular appeal in the international arena because it appears to offer a neutral basis for reaching decisions on issues that are otherwise mired in controversy. A more critical view of the use of science in policymaking, however, reveals that the science itself is the product of its social and political context. Thus the "epistemic communities" that appear to offer a coherent scientific understanding of our environmental problems can be seen, at least, as influenced by domestic political interests, or, more boldly, as the product of a hegemonic world order in which the local knowledge of peasants and indigenous peoples is just as marginalized as their political power.

WHO HAS STANDING?

The common thread running through all these arguments is: Who counts? Up to now the discourse on global environmental change has been the province of state-based actors and a narrow group of experts. These are now being challenged by the groups who are most directly affected by environmental change and who will bear the brunt of government policies. For the most part these groups have not had a place at the table and yet their claims have legitimate moral authority. At the same time, they must recognize that there are other legitimate claims to be considered. Where there are competing universalistic claims, we are entitled to examine the particularistic claims from which they arise.

It is sometimes argued that social justice can only be served by broad access to and actual participation in decision making. Without such broad participation, so the argument runs, we cannot be sure that all legitimate interests have been voiced and given standing. We agree with this principle, but not unconditionally. First, participation alone does not ensure equality. Some interests are more equal than others because not all have the resources to mobilize or enjoy equal access to power. Second, the role of nongovernmental organizations in promoting the interests of hitherto unrepresented points of view cannot be assumed. They can be coopted; they may have narrow bases of support; and some of their actions may be based on principles that challenge international law. Third, there is simply the problem that not everybody can sit at the table (including unborn generations). How should those who are not at the table be represented? Fourth, it should be recognized that a too inclusive approach to global problems may simply be a prescription for paralysis. Kenneth Arrow's impossibility theorem that democratically based solutions are suboptimal and may, in the end, satisfy no one is a very real dilemma in global ecopolitics.

Even in the absence of full participation, however, we can still insist that moral arguments be an explicit part of the debate. Although we cannot resolve many specific issues—such as, for example, the competing land claims in Amazonia—we can insist that clearly articulated ethical principles inform policy choices. Claims for authority, whether knowledge-based, rooted in state structures, or identified with substate actors, must be examined for their implications for norms of social justice. Proposed policies must be evaluated for their effect on all relevant groups and on the survivability of the ecosystem. Taken seriously, such a charge would transform the negotiations over appropriate responses to global environmental change. It would be a first step toward establishing criteria for steering the response to environmental problems so that the human race does not continue to perpetuate those practices and behaviors that led to the problems in the first place.

I

Environmental Change and the Varieties of Justice

HENRY SHUE

PREFERENCE AND FAIRNESS

THE various policy communities, including those focused on the environment, tend to accept existing human preferences uncritically when making social choices. Policy analysts seem to do everything but challenge these preferences. They adopt them as if they were a baby left on their doorstep.[1] If, for instance, many people in metropolitan areas happen to prefer traveling to work in private automobiles rather than in commuter trains, policy analysts will judge the health of the economy by the ease with which these people obtain and operate the cars they want. Yet there is no good reason to treat the preferences held at any one time, including the present time, as if they were somehow foundational and immune to assessment when we are setting long-term public policy. Indeed, total passivity toward all current human preferences is hopelessly misguided, especially in the face of environmental change caused by global warming. We must decide which preferences to modify in this particular case, however difficult it may be to bring about such change either in ourselves or in others.[2]

---

[1] This point has been powerfully established over the years by Mark Sagoff. See his *Economy of the Earth: Philosophy, Law, and the Environment* (Cambridge: Cambridge University Press, 1988), especially chap. 5, "Values and Preferences."

[2] It is possible that Americans are more ready to consider change than my worst fears suggest. "People are struggling with deep ambivalence about their own values. . . . Watch television for a day and you will get a clear picture of what Americans supposedly want in life: new cars, a big house, stylish clothes, the latest gadgets—and of course, fresh breath. Yet when Americans describe what they are looking for in life,

Most preferences entail a claim upon resources.[3] The preference for operating private automobiles, for example, entails claims upon, among many other things, the energy and the raw materials needed to make the necessary steel, the gasoline needed to operate the combustion engine, the hospitals needed to treat the injuries from the accidents, and the capacity of the planet somehow to deal with the tons of carbon dioxide emitted annually by each car, along with all the other troublesome—and toxic—emissions. Because the satisfaction of a preference ordinarily demands the consumption of resources (and in this case, the creation of emissions), to calculate from existing preferences is in effect to postulate that all (and only) current preferences are created equal. One is tacitly assuming without explanation that no matter how divergent the respective "calls" of two different preferences upon a society's or the planet's resources, the preferences are, at least prima facie, equally reasonable and equally urgent.

This is manifestly implausible. One person's desire for an additional jar of caviar is not equal in urgency to another person's need for an additional bowl of black beans. One state's intention to spend three hundred billion on military preparations is not equal to another's need to spend three billion on infrastructural improvements. Yet if we fail to question the relative value of such competing preferences, then we ask only, Does the party in question in fact command enough credit to borrow the necessary sum? We fail to ask, Is this a preference that deserves to be honored, or even tolerated, by the rest of us (including, presumably, future generations who will in due course have preferences of their own). We fail to ask, Will the satisfaction of this preference consume limited resources that should be used differently—or protected from use, either now or forever?

Whenever I, my state, or my group wants something, this usually means that we propose to consume whatever social wealth, natural resources, political attention or indulgence, and individual time and energy needed to produce it. No matter how productive the project, it will still require resources that could have been used many other ways, or preserved in their current state and not consumed for human purposes at all. Everyone understands that investment and consumption

---

their aspirations rarely center on material goods." *Yearning for Balance: Views of Americans on Consumption, Materialism, and the Environment* (Bethesda, Md.: Harwood Group, 1995), p. 14. Four focus groups are reported on.

[3] This is lucidly noted by Will Kymlicka, *Liberalism, Community, and Culture* (New York: Oxford University Press, 1989), pp. 37–38. Obviously the limiting case is a preference for a service that can be performed without consuming any resource or expending any marketed energy.

have opportunity costs, but there is more at stake here than the obvious costs. Why should just *these* preferences control *these* resources? Whose preferences control policy? Who pays the costs? Who receives the benefits? Whose costs and benefits "count," and whose do not? *Every* calculation presupposes such a division between those internal to the group and those external to it.[4] Those whose preferences count, according to the method of social choice used, control the resources. One cannot think seriously about any environmental policy without thinking seriously about which preferences should be taken into account in setting the policy. "Ours now" may seem the obvious answer to us now, but it will not seem obvious to the people of distant places and distant times.

To decide whose preferences count, we must first examine our grounds for the selection of preferences. To assume that any party—individual, group, or institution—is entitled to satisfy all and only the preferences that it is in a position to satisfy without *in future* engaging in force or fraud is to assume that no reasonable questions can be raised about past or present holdings of resources. Iain Wallace and David B. Knight make such an assumption near the beginning of Chapter 4 in this volume. After defining the net primary organic productivity (NPP) of a regional ecosystem as "essentially its capacity to convert incoming solar energy . . . into terrestrial vegetation, on which 'higher' lifeforms ultimately depend," Wallace and Knight conclude with an ethical judgment:

> By this measure, a tropical rain forest is intrinsically more productive (has a higher mean NPP) than a boreal forest, a high-altitude grassland, or a hot or cold (tundra) desert, for instance. Other things being equal, a "traditional" or preindustrial society situated in an ecosystem of higher NPP will enjoy a higher sustained material standard of living than one situated in an ecosystem of lower NPP, as a direct result of the underlying non-isotropic natural conditions. Economic *inequality* between societies in these circumstances is not, prima facie, an *inequity*.

This seems to assume, in general, that human history counts for nothing and, in particular, that all present territorial holdings are equally beyond challenge—that each people has precisely the territory and resources it ought to have—however much force, fraud, or environmental havoc lie behind the current locations of borders, resources, and

---

[4] This is true whether the calculation concerns some kind of subjective preferences or some kind of objective costs not calculated on the basis of preferences. I leave the much-discussed subjective/objective problem aside in favor of the less-noticed internal/external problem, which comes up either way.

persons. Stated clearly, these assumptions are difficult to credit. That many children, through no fault of their own, are born into regions that simply cannot sustain a high standard of living no matter how hard people try, whereas others, through no merit of their own, are born into regions that do sustain a high standard of living throughout life is prima facie an inequity, an extreme inequality in life prospects that cries out for either change or justification. Why should the life prospects of two otherwise similar infants be so radically different for reasons entirely beyond their control?

The only alternative to denying the significance of the contingencies, crimes, and accidents in the history of the world so far is to construct some kind of reasonable standards (other than the mere fact of possession) for the assessment of current holdings of territory and other resources. This involves the formulation of what philosophers call principles of distributive justice. If it is incredible that the status quo must simply be accepted as uniformly justifiable whatever the divergences in the undeserved fates of individuals, we must try to establish some criteria for distinguishing the outrageous from the tolerable. We must try to make sense of our conviction that not everything is equally acceptable by spelling out some standards of acceptability in the division of resources, including most especially the initial allocation from which our markets and other distributive processes must begin. And we have to do this before we can know *whose* actual preferences, and *which* of them, to count.

Still, how much do we need to know about justice in order to make environmental policy? If several parties (individuals or groups) have conflicting preferences, they would do well to work out some mutually acceptable arrangement. Yet don't they need to have a set of principles of justice before they can arrive at a limited plan of action?

If parties are more or less equally situated, they should bargain directly. Other things being equal, it may be best if parties can work out among themselves the terms of any dealings they will have with each other. Even lawyers, however, have the concept of an unconscionable agreement; and nonlawyers have no difficulty seeing that an agreement can have objectionable terms if one party is subjected to some kind of undue influence. Parties can be unacceptably vulnerable to other parties in more than one way, naturally, but perhaps the clearest case is extreme inequality in initial positions. This commonsense understanding means that even morally acceptable bargains depend upon initial holdings that are not morally unacceptable.

Obviously to recognize acceptable bargaining we must have some knowledge, at least, of the standards for fair shares. If we do not know whether the actual shares that parties currently hold are fair, we cannot know whether any actual agreement they reach is morally conscionable. The simple fact that they all agreed is never enough. They don't have to have a complete theory of justice before they can agree upon practical plans, but they do need to know the relevant criteria for fair shares of holdings before they can be confident that any plan they actually work out should in any way constrain those who might have preferred different plans. As Chris Reus-Smit forcefully reminds us in Chapter 5, there are deep issues of justice in addition to issues about fair shares, but here I can sketch only some of the latter.

FAIRNESS AND GLOBAL WARMING

It is time to concentrate on a concrete instance of environmental change: climate change, or "global warming." In general, issues of fairness fall along multiple continua: (1) fairness specific to the problem at hand versus general background fairness; (2) fault-based standards versus no-fault standards; and (3) standards for transitions (or extrications) versus final standards. In sum, conceivably either the fairness of background conditions or the fairness in handling a specific problem might be judged by either fault-based or no-fault standards that are taken to be either transitional or final. Since each of these three kinds of considerations are presumably continuous, comments on fairness can be located in many different three-dimensionally specified conceptual locations. Quite a few disputes consist, not over different answers to the same question, but over questions not obviously different. Here we need to see the precise shape that various issues of fairness take in the case of a particular kind of environmental change, global warming.[5]

The basic issue of fairness characteristic of global warming seems to be: How should responsibility for solving the problem of global warming be divided, given how responsibility for creating the problem is in fact divided? And given how the benefits of the activities that produce the problem have in fact been distributed? The most important contributors to global warming are gases produced by the industrial activities that have made the rich rich. How much should the

[5] A comprehensive view of the issues specific to global warming, splendidly integrating economic and ethical considerations, appeared since this writing: Michael Grubb, "Seeking Fair Weather: Ethics and the International Debate on Climate Change," *International Affairs* 71 (1995): 463–96. Also see Matthew Paterson, "International Justice and Global Warming," in *The Ethical Dimensions of Global Change*, ed. Barry Holden (London: Macmillan, 1996).

poor contribute to the solution of a problem caused by activities from which they have not benefited nearly as much? The division of causal responsibility for global warming and the division of benefits from the activities causing global warming are conceptually distinct (even if in fact they tend to fall together) but both are specific to the problem at hand.

A little further reflection suggests that "the problem" is at least two distinct problems: (1) What is a fair allocation of the costs of avoiding however much additional warming still can, and reasonably should be, avoided? And (2) What is a fair allocation of the costs of social adjustments to however much warming is not in fact avoided? The answers may be similar; I have tried elsewhere to show that they are profoundly interconnected.[6] Yet no matter how interconnected, these two questions about global warming are still distinct.

One might also pursue a *general* inquiry: How should responsibility for solving the problem of global warming be divided in light of the background inequalities in wealth and power that are the present bitter fruit of centuries of colonialism, imperialism, unequal development, war, greed, stupidity, or whatever exactly one thinks are the main features of the history of the international political economy? One might think that background injustices not directly connected to the origins or fruits of global warming itself dwarf in significance any injustices that are. Or the opposite—that is part of what would have to be considered in analyzing the problem in depth, as I would try to do if I were here not merely outlining the normative agenda.

I establish elsewhere that the gross inequalities among nations in wealth and power, and therefore in bargaining leverage, will undercut the moral legitimacy of the outcomes of actual bargaining among nations on the matter of global warming.[7] Actually, as long ago as Stockholm 1972, the rich countries conceded the principle that the poorer countries are entitled to special arrangements that they are in no bargaining position to demand; one understanding of this was reaffirmed in the terms of the Montreal Protocol (1987) on ozone destroyers and again in Principle 6 of the Rio Declaration on Environment and Development (1992), which begins: "The special situation and needs of de-

---

[6] Henry Shue, "The Unavoidability of Justice," in *The International Politics of the Environment: Actors, Interests, and Institutions*, ed. Andrew Hurrell and Benedict Kingsbury (New York: Oxford University Press, 1992), pp. 373–97, especially (on the practical inseparability of these two issues) pp. 384–97.

[7] Ibid. For a contrasting approach, see Adam L. Ironstone, "From 'Cooperator's Loss' to Cooperative Gain: Negotiating Greenhouse Gas Abatement: Note," *Yale Law Journal* 102 (1993): 2143–74.

veloping countries, particularly the least developed and those most environmentally vulnerable, shall be given special priority."[8]

Some standards of fairness refer to, and therefore depend upon an accurate analysis of, the respective causal roles already played by the various parties.[9] Notably, portions of Principle 7 of the Rio Declaration contain the following rather delicate (and exceedingly vague) statements: "In view of the different contributions to global environmental degradation, States have common but differentiated responsibilities. The developed countries acknowledge the responsibility that they bear in the international pursuit of sustainable development in view of the pressures their societies place on the global environment."

On the other hand, a different standard of fairness refers strictly to features of the existing situation without any reference to their origins or development. Again, the remainder of Principle 7 of the Rio Declaration states ". . . and [in view of] the technologies and financial resources they command." Such payment according to ability to pay, for example, is a no-fault standard. It in no way presupposes or implies that the better-off did (or did not do) anything wrong in the process of becoming better off. This rationale for assigning heavier burdens to the better-off refers forward, not backward; it concerns how well the better-off can bear the heavier burdens, not how they came to be better off.

The fault-based "polluter pays principle" (already established within the OECD and embraced explicitly, with qualifications, in Rio Declaration, Principle 16), by contrast, assigns heavier burdens to those who have contributed more heavily to the problem (regardless of ability to bear the burdens). Obviously one notable way by which parties come to have greater ability to pay is by having externalized the costs of their own gains: by not paying for the effects of their own pollution. In such cases, ability-to-pay and polluter-pays may select the same parties, but that is a contingent conjunction in the applications of principles that are conceptually distinct. These two principles may assign different degrees of responsibility to any one party, making

[8] The Rio Declaration is widely reprinted; one source is *Agenda 21: The United Nations Programme of Action from Rio* [Final Texts of *Agenda 21*, Rio Declaration, and Statement of Forest Principles, E.93.1.11] (New York: Department of Public Information, United Nations, 1993), pp. 9–11. A valuable commentary on possible precedents in the Montreal Protocol is Jason M. Patlis, "The Multilateral Fund of the Montreal Protocol: A Prototype for Financial Mechanisms in Protecting the Global Environment," *Cornell International Law Journal* 25 (1992): 181–230.

[9] As just indicated, the causal role investigated could be either specific to global warming or general to the world economy.

the Rio Declaration ambiguous in principle, which may or may not matter in practice.

The application of a no-fault standard depends upon the facts about the current situation—such as the extent of inequalities in standards of living and technological resources—whereas the application of a fault-based standard depends upon facts about stretches of the past, often rather long and wide stretches. Consequently, attempts to apply fault-based standards are virtually guaranteed to become embroiled in more or less irresolvable controversy about historical explanations. Yet never to attempt to assess fault is to act as if the world began yesterday.

Philosophers and political theorists have written little about what might be called principles for transitions, either principles of the fairness of transitions or principles of any other kind directly applicable to how change is brought about. Normally, we are offered an ultimate ideal and, in effect, wished good luck in figuring out how to reach it. I think this is lazy and irresponsible, but transitions are hard, intellectually as well as practically. Consider "extrication ethics": here the central question is not simply, How do we manage to reach the ideal, but How do we manage to reach the ideal given that we are already in a swamp, as the saying goes, "up to our ass in alligators"?[10] Since I am inclined to think the swamp general and the alligators numerous, I tend to think that extrication ethics largely constitutes transition ethics—but no matter. Wherever we are, we face questions about how to get to where we are supposed to be, as well as questions about where exactly that is.

One of my teachers used to talk about what he called the pacifist's fallacy. Since I think pacifism is quite a reasonable position—although I do not hold it—this strikes me as an unfortunate name. Nonetheless, the underlying point is important. And the point is that even if we know beyond all doubt that the ideal society is one without violence, it simply does not follow that we should start being nonviolent now and count on the power of our example to bring others around to our way of thinking. It may be that when we consider exactly where we are starting from—for example, a "swamp" of injustice, unemployment, exploitation, and looting—we need to begin by applying state

---

[10] I take the term "extrication ethics" from Tony Coady, who has written one of the few helpful discussions of the subject of which I am aware. See C. A. J. Coady, "Escaping from the Bomb: Immoral Deterrence and the Problem of Extrication," in *Nuclear Deterrence and Moral Restraint*, ed. Henry Shue, Cambridge Studies in Philosophy and Public Policy (New York: Cambridge University Press, 1989), chap. 4, especially pp. 193–225.

violence to end popular violence, or popular violence to end state violence. (Or the pacifist may actually be correct that violence from either side will only lead sooner or later to more violence, and that we would do best to start being nonviolent right away.)

The point is just that even if the pacifist should turn out ultimately to have been correct, she will need to have given some kind of means/end analysis in order to show that she is correct.[11] The rationale for any transition/extrication ethic needs to be spelled out well beyond a statement of the ideal. The transition ethic *may* be the same as the ideal ethic: sometimes the best route to the ideal is to set an uncompromising example of the ideal, right here in the swamp. Since, in general, much human activity is self-defeating, we cannot, however, simply assume that good examples will routinely inspire emulation, whatever the circumstances.

In sum, then, the minimum list of essential, and unavoidable, questions about distributive justice in the case of global warming contains at least the following:

1. What is a fair allocation of the costs of preventing the global warming that is still avoidable?
2. What is a fair allocation of the costs of coping with the social consequences of the global warming that will not in fact be avoided?
3. What background allocation of wealth would allow international bargaining—about issues like (1) and (2)—to be a fair process?
4. What is a fair allocation of emissions of greenhouse gases (a) over the long term and (b) during the transition to the long-term allocation?[12]

[11] Philosophers will be quick to note that this assumes that the agent is to some extent responsible to advance, or at the very least, not to retard, the ideal and not simply to embody it, whatever the consequences. I am sailing over deep waters here but hope not to have to plunge into them.

[12] These questions are explored at greater length in Henry Shue, "Subsistence Emissions and Luxury Emissions," *Law & Policy* 15 (1993): 40. The heart of this analysis has also appeared as Henry Shue, "Four Questions of Justice," in *Agricultural Dimensions of Global Climate Change*, ed. Harry Kaiser and Thomas Drennen (Delray Beach, Fla.: St. Lucie Press, 1993), pp. 214–28. Responding to the version of this analysis into four conceptually separable questions in *Law & Policy*, Michael Grubb has emphasized the depth of the economic ties between allocation of emissions (4) and allocation of the costs of abatement (1). See Grubb, "Seeking Fair Weather," especially pp. 483–88. If a nation's allotment of permissible emissions is well below what would have been its business-as-usual level (not to mention its current level), the nation will need, other things being equal, to spend large amounts on abatement, either its own or other nations', perhaps through "joint implementation." Allocating the costs of prevention in

At least these four issues need to be explored in order to develop reasonable policies. In the remainder of this chapter I struggle a bit more only with (4b), which I find the hardest to grasp in the case of global warming, as an illustration of what might be involved in taking fairness seriously in one instance of environmental change.

COMING FULL CIRCLE

The underlying question remains: Who counts? It seems to me—although nothing about this is entirely clear—that the answer has to be, roughly, everyone whose fundamental interests are seriously affected. There are fewer and fewer tolerable environmental externalities, because there is no one "out there" on whom the significant externalities may be inflicted. The internal/external line must be drawn, if it can be drawn at all, where the actual serious consequences end—the consequences that affect fundamental interests. *Whose* fundamental interests? Anyone's that are seriously affected. I call this the naturalistic approach: one follows the effects of the activity in question at least as far as they continue to impinge significantly on human beings (and on ecological elements that play an irreplaceable role in human life).[13] If the activity is, say, setting the discount rate for U.S. banks, and the effects include jobs in Mexico, then, assuming that jobs are vital resources for living any kind of acceptable life, and that resources for living individual lives are the interest at stake, the justice of a particular discount rate and/or a particular process for setting discount rates depends in part on its effects on Mexican jobs—not just on its effects on U.S. jobs.

The alternative approach can be called voluntaristic, because it depends upon the implicit contention that one can somehow renounce responsibility for the significant effects one is in fact having upon some people. Keeping the same example, one would say, on the voluntaristic approach, "Yes, I realize that this is having significant negative effects on Mexican jobs, but my social contract is exclusively with U.S. workers and so I do not take effects on Mexican jobs into account." This might be part of a theory about the obligations of citizens to one another or of a theory about the obligations of a government to the

order to produce an allocation of emissions that had been chosen independently on grounds of fairness would of course be one no-fault allocation of prevention costs. The answer to question 1 would be derived from an answer to question 4a.

[13] As fewer and fewer people remain on the external side of the line, the tenability of the internal/external line itself is jeopardized, as was noted by an anonymous reviewer, who also emphasized the ecological elements irreplaceable in human life.

governed. Its crucial feature is that effects on some people, the out-siders, may be disregarded—or discounted.

Discounting outsiders is, nowadays, more plausible than totally ig-noring them. Perhaps it would not be acceptable to give the outsiders' interests no weight at all, but they are to be given significantly less weight than the interests of insiders. What seems crucial (and strange) about this—and the reason why I call it voluntaristic—is the assump-tion that it is up to the insiders to *decide* how much, or whether, to weigh the interests of outsiders in their balance. I have only the re-sponsibilities I choose to take upon myself.[14] We have only the respon-sibilities we choose to acknowledge. What spectacular assertions of autonomy! Could anything be less compatible with an understanding of ecological interdependencies?

What if I begin to smoke a cigar during a meeting, and people begin to complain. But the complaining people are all women, and I say: "I decided long ago not to acknowledge any responsibility toward women"? What if we build a factory that discharges pollutants into the Rhine in Switzerland, and people begin to complain. But the com-plaining people are all Dutch, and we say: "We decided long ago not to acknowledge any responsibility toward non-Swiss"? What if we build an economy that lays debts upon future generations, and [self-appointed spokes-]people begin to complain. But the people com-plained on behalf of are all nonexistent, and we say: "We decided long ago not to acknowledge any responsibility toward nonexistent peo-ple"?

Whatever else can be said about these three cases (and the differ-ences among them), it is preposterous to think that I, or a self-defining "we," can simply decide as we please whether we fancy taking on responsibilities, construed as optional, toward people defined by us as others or outsiders, when their lives are being affected by ours. Volun-tarism does not plausibly extend this far.

Moral responsibility for correcting injustice must, to a considerable extent, track causal responsibility. This ties it to what I call fault-based standards; this is a very specific way in which the topics of relevant communities and justice interlock conceptually. When my pursuit of my interests affects the resources available to others for their pursuit of their interests, I cannot *simply* declare those others out of bounds: "Sorry, but you are the wrong sex, or nationality, or generation to count under my system." It may be highly relevant whether I inter-

---

[14] This level of would-be voluntarism has been subjected to a strong critique in Nancy J. Hirschmann, *Rethinking Obligation: A Feminist Method for Political Theory* (Ithaca, N.Y.: Cornell University Press, 1992).

fered with your pursuit of your interests intentionally, or accidentally, or unknowingly, or unforeseeably, or how much harm my own interests would have suffered if I had chosen to protect yours, and so on. But it is quite irrelevant that I would simply prefer not to count you and your kind at all—that cannot be simply for me to say, especially if shortages or harms I produced, however unintentional and unforeseen, deeply affect the livability and richness of the only life you can expect to have. The voluntaristic view is like the renunciation of paternity: renounce it as one may, the objective fact of one's effect on the world remains. One can deny responsibility, but one cannot remove it.

What, then, should we be doing in the short term (in the initial transition away from where we are now) specifically about global warming?

TRANSITION/EXTRICATION

Change is rarely instantaneous, and we must, therefore, recognize in most cases some kind of transition period between wherever we are now and where we ought to be. But there are transitions and transitions—it matters very much how long a transition takes, and it especially matters whether some people live their entire lives and die before the transition is completed. If there is anything to which every human being has a right, it is no good for the people whose lives have meanwhile come and gone that arrangements are gradually being made so that eventually everyone will enjoy what those now dead were also entitled to have enjoyed but did not.

Let me focus on two rough categories of ethical standards, which I will call minimum standards and short-range standards. These are not traditional labels, and there is nothing sacred about these particular names for the categories, which are meant to be one extreme and the middle of a continuous scale. Obviously the other extreme on this continuum could be called long-range standards. The next item in that direction, beyond long-range standards and off the ethical charts, is something like high-minded hypocrisy: empty talk about what would be "great" without any sense of obligation to act. Thus by "long-range standards" I do not mean dreamy rhetoric; I mean genuine standards that we are bound to meet as soon as we reasonably can, but which are in fact going to take a while to satisfy even if we pursue them with appropriate vigor. For example, many of us believe that we ought to eliminate racist and sexist attitudes as well as racist and sexist behavior. But no matter how aggressively we attack racist and sexist behavior, there is no way that we can in the short run eliminate the attitudes that lie behind the behavior. Besides eliminating the behavior, we can also criticize and pour contempt upon the attitudes, but that will not destroy the attitudes, in ourselves or in others. This persistence of the attitudes is definitely not a minor problem that we can happily live with. The trouble is that the only way to try to eliminate the recalcitrant attitudes immediately would be to engage in some kind of "moral reeducation" that would look a lot like brainwashing, and that would be either ineffective or highly undesirable in light of some of our other firmly held ethical standards. So long-range standards specify, not places to which "you cannot get from here," but places we cannot go straight to right away.

I have already by implication given an example of what I take to be a short-range standard. There are reasons why we cannot suddenly make all men respect women—why we cannot change all our attitudes—but we can make men stop harassing women. I am not of course predicting that we can in fact be 100 percent successful—nothing is 100 percent successful—but we can seriously and soberly adopt a policy that says, not "We want to get sexual harassment down to tolerable levels" (whatever "tolerable levels" might mean), but "We intend to stamp out sexual harassment." Even this cannot be done tomorrow, of course, which is why it is a short-range and not an "instantaneous" standard. Laws will have to be passed against harassment, regulations will have to be issued defining it, victims will have to find the courage to make accusations even though they know they will be considered troublemakers, accused men will have to have fair hearings, convicted men will have to be punished, and word will have to get around that harassing women is just not much fun anymore. This will all take some time, but it need not take an awful lot of time, particularly if the attitudes also begin to change, so that the men do not just figure that it is not worth the hassle any more but come to believe that women deserve better.

Beyond short-range standards lie what I call "minimum standards." Since the other categories are called "long-range" and "short-range," it would have been natural to call this one "immediate," and by "minimum standards" I do mean standards that must be satisfied as quickly as is humanly possible. These minimum standards must always be fully satisfied. In principle, no transition period, however brief, is acceptable. As always, we will probably not in fact succeed in effecting such standards immediately, but we should try, sincerely and seriously. In my analogies of racism and sexism, now we are talking about lynchings and rapes. If we said to someone, "Look, lynching—or rape—is just unacceptable, and we want it stopped," and he said, "Right—I've got a plan, and we will reduce lynchings to 90 percent in one year, to 80 percent after two years, and so on until in a decade we will have come close to wiping out lynching," I think our response would be, "You don't seem to understand about the seriousness of lynching—we don't want no lynchings by the year 2006—we want no lynchings now."

The practical difference may be best seen in budgetary terms. Suppose (unrealistically) that it turned out that for every 10 percent increase in the enforcement budget of the FBI you could cut lynchings by 10 percent, and someone proposed that the budget should go up 10 percent per year for 10 years, I think the natural response would be: "No—we want the budget for this increased 100 percent immediately because we can't think of anything worse than lynching that merits the money more."

The budgetary analogy suggests another useful way to mark the differences among long-range standards, short-range standards, and minimum standards, namely, the amount of sacrifice one should be willing to make in order to satisfy the standard. Once again, this is a continuum, not a series of bright lines, but the differences are significant. For a long-range standard one is serious and is making some effort, but one is also working on many other fronts, some of which are more urgent. For a short-range standard one is making serious investments of time, money, and energy, although a very few things are more urgent still. Those "few things [that] are more urgent still" are the minimum standards, which have higher priority than anything else, with the possible exception of any necessary preconditions of them all, such as the prevention of the extinction of the human race. At least some of the requirements of justice are, I think, minimum standards.

Unfortunately, one cannot specify what we should do in the short run without having some answer to where we should arrive in the long run; one needs to know the objective before one can concretely consider whether it may be compromised. In terms of my earlier analysis of four questions of distributive justice, one cannot answer (4b) independently of (4a). Discussions so far of the allocation of greenhouse-gas emissions that would be fair over the long run have tended toward the simplistic extremes. At the one extreme, those with rich-country interests in mind tend to assume that current emissions constitute some kind of unquestionable baseline of entitlement (from which we should simply start trading emissions permits), as if the great Hegelian fantasy had spontaneously materialized: the real is rational and just, and the rational and just is real. That there is some type of property right in current levels of emissions is an intelligible thesis, but one that is far from obvious (and, I argue, far from true).

At the other extreme, those with poor-country needs in mind tend to assume that if one grants that all human beings are equal in dignity and worth, it immediately follows that all human beings have a right to equal greenhouse-gas emissions, which amounts, roughly, to equal levels of economic activity. Anil Agarwal and Sunita Narain famously endorsed equal per capita emissions in their much-cited (but, I suspect, less read) 1991 monograph; and Dr. Atiq Rahman, director of the Bangladesh Center for Advanced Studies, said the same in August 1993 before a hearing of the U.S. House of Representatives.[15] While I believe these advocates of equal emissions are much closer to the truth than the defenders of current emissions, I do not think that either polar position can be adequately supported.[16] This turns out to be a fascinating but highly complex issue, which I cannot begin to analyze here. Elsewhere I have defended the view that the universal human right is to, not equal emissions *simpliciter*, but an equal minimum; if ever the minimum were universally guaranteed, departures from equality could then be justifiable.[17] While the difference between strict overall equality and an equal minimum could in the end be significant, in the transition—which is to say, now—it is insignificant.

I take it, then, that our goal is to allow every human being an equal minimum level of greenhouse-gas emissions.[18] We can even assume for the sake of this argument, in order to make things easier for ourselves, that the minimum level is quite low, roughly, subsistence. What can we say about the transition period that starts now? Since this is, strictly speaking, a minimum standard, its implementation may not be compromised or delayed. We should immediately see to it that no one is ever denied her minimum share of emissions. What does this mean concretely?

[15] See Anil Agarwal and Sunita Narain, *Global Warming in an Unequal World: A Case of Environmental Colonialism* (New Delhi: Centre for Science and Environment, 1991), and "Bangladesh Environmentalist Calls for Division of Global Emissions Allowances," *International Environment Reporter: Current Reports* 16 (1993): 574.

[16] For a pioneering attempt to deal with the ethics of transition in the case of climate change by a progressively changing weighing of current emissions and population, see Michael Grubb, James Sebenius, Antonio Magalhaes, and Susan Subak, "Sharing the Burden," in *Confronting Climate Change: Risks, Implications, and Responses*, ed. Irving M. Mintzer (Cambridge: Cambridge University Press for the Stockholm Environment Institute, 1992), pp. 305–22 (for the formula, see note 7 on p. 321).

[17] Henry Shue, "Avoidable Necessity: Global Warming, International Fairness, and Alternative Energy," in *Theory and Practice*, NOMOS 37, ed. Ian Shapiro and Judith Wagner DeCew (New York: New York University Press, 1995), pp. 239–64.

[18] This is my answer to question (4a). I have developed other implications of this answer in Henry Shue, "After You: May Action by the Rich Be Contingent upon Action by the Poor?" *Indiana Journal of Global Legal Studies* 1 (1994): 343–66.

Here is where the concrete problem begins to get interesting. A broad consensus of scientists tells us that the global *total* of greenhouse-gas emissions must be *reduced* below the 1990 total if serious global warming is to be avoided—emissions are not, I emphasize, to be stabilized *at* the 1990 level (which was the Clinton administration's great new commitment as of Earth Day 1993) but to be reduced well *below* it.[19] The original report of the Intergovernmental Panel on Climate Change (IPCC) said that CO_2 emissions have to be reduced 60 percent below the 1990 total! If a political commitment is made to act upon that scientific consensus, the total of emissions becomes a *shrinking* zero sum: the room under the global ceiling for any additional emissions to sustain a newborn child, or to move a malnourished child back from the brink of irreparable physical damage, must be created by reductions in someone else's current emissions, and then significant further reductions must be made in current emissions in order to make the total shrink. Everyone understands that the waste of fossil fuels in most rich countries is so egregious that there are "no-regrets" reductions in fossil-fuel energy consumption to be made—the initial cuts in carbon emissions would be not merely painless but profitable.[20] Nevertheless, this provides only a kind of grace period before the crunch comes.

The problem remains dramatic even if we wave a magic wand again and make it much easier still. Assume, to ease the strain, that the scientific consensus, as originally reported by the IPCC, is wildly alarmist and that we do not need to go below 1990 levels but need only to maintain 1990 levels. Perhaps, to continue the fairy tale, the elimination of waste in the rich countries would make enough space under the global ceiling (constituted by the 1990 total) to accommodate subsistence levels of economic activity by everyone currently liv-

[19] See J. T. Houghton, G. J. Jenkins, and J. J. Ephraums, eds., *Climate Change: The IPCC Scientific Assessment* (New York: Cambridge University Press for the Intergovernmental Panel on Climate Change [WMO/UNEP], 1990); the brief "Policymakers' Summary" is also available in a separate and more colorful form as *Scientific Assessment of Climate Change: The Policymakers' Summary of the Report of Working Group I to the Intergovernmental Panel on Climate Change* (Geneva: Intergovernmental Panel on Climate Change [WMO/UNEP], 1990). Also see J. T. Houghton, B. A. Callander, and S. K. Varney, eds., *Climate Change 1992: The Supplementary Report to the IPCC Scientific Assessment* (New York: Cambridge University Press, for the Intergovernmental Panel on Climate Change, 1992). For the Clinton Earth Day speech, see "Reaffirming the U.S. Commitment to Protect Global Environment," *U.S. Department of State Dispatch* 4 (April 26, 1993): 277–80 (reduction to 1990 levels by 2000: 278, col. 3).

[20] Alliance to Save Energy et al., *America's Energy Choices: Investing in a Strong Economy and a Clean Environment* (Cambridge, Mass.: Union of Concerned Scientists, 1991), p. 2; Michael Shepard, "How to Improve Energy Efficiency," *Issues in Science and Technology* 7 (Summer 1991): 85–91.

ing in absolute poverty. Still, even with a few such fantastic favorable assumptions, it is clear that, given current fuels for economic activity, sooner rather than later *people in rich countries must voluntarily and indefinitely reduce their level of economic activity* in order to allow for additions to the human population, who share in the universal human right to a minimum level of economic activity. Even with all my magical weakening assumptions, we may be contemplating revolutionary change.[21]

Here are the primary options, explicit and implicit (civilized and barbaric):

1. We can simply renounce any shred of commitment to human equality, abandoning even any universal right to an equal minimum, and instead acknowledge that in our hearts we actually do think that there are two kinds of people, people worthy of life and people not necessarily worthy of life, with the "preferences" of the latter to be excluded from our calculations if need be.

2. We can acknowledge the scientific consensus about what is likely to happen if we do not adopt mitigation strategies now, but opt nevertheless for a policy favoring adaptation later over mitigation now and thus not requiring the capping of total emissions, in short, deny that the science justifies the policy of mitigation.

3. We can voluntarily and indefinitely reduce our level of economic activity in order to reduce emissions.

Or 4. we can create energy technology that does not produce (as much) greenhouse-gas emissions.

As far as I can see, these are the most eligible choices (and I would be delighted to be shown another good one).

The first option, I hope, is out of the question. Until the specters of fascism and apartheid become too dim any longer to haunt us, we are, I hope, emotionally incapable of abandoning all pretense to a commitment to some minimal equality across humanity, however little exertion we expend in implementing our rhetorical stand and however tortured some of our interpretations of what equality means. The equanimity with which Western Europe and the United States are tuning out the "ethnic cleansing" of the Bosnian Muslims in the former Yugoslavia is, however, not encouraging. Still, if we can be optimistic about our commitment to equality, we effectively have three options: (2), (3), and (4).

Unlike option 1, option 2 (denying that the policy of immediate

---

[21] And I am for now leaving aside entirely a host of possibly compounding problems about the interests of nonhumans rightly raised by Wendy Donner in Chapter 3.

steps toward mitigation of global warming is well enough supported
by the science) strikes me as a perfectly sane *type* of argument. Since
scientific findings alone can never dictate what we ought to do but can
only give us (the absolutely vital) information about the paths that are
physically possible and the comparative risks and benefits of the alternative paths, it is always in principle perfectly reasonable to challenge
any policy prescription based explicitly only on the science, since implicit values are necessarily being assumed where they are not explicitly invoked.[22] Nevertheless, in this particular instance, the arguments
are quite weak for continuing to increase the rate at which we inject
greenhouse gases into the atmosphere in spite of the magnitude of
uncontrolled and unpredictable climate change that is clearly possible
and that many scientists think is virtually inevitable.

The strongest case I have seen for later adaptation instead of present
mitigation is constructed by Oxford economist Wilfrid Beckerman. He
advocates continuing to accumulate wealth through productive investment of the resources that might otherwise be dedicated to mitigation
and relying later upon the additional wealth thereby accumulated to
pay for whatever specific adaptation is required by the actual effects of
whatever warming does in fact occur.[23] One fatal flaw in this proposal
is that those who seem most likely to have their lives threatened by the
effects of warming—for example, Bangladeshis displaced by rising sea
levels—are precisely those least likely to have accumulated the resources to pay for adaptation if no measures have been taken in the
meantime to achieve that end. That the worst risks fall upon the most
vulnerable is especially clear if Cynthia Rosenzweig and Martin L.
Parry are correct that although temperature changes may be smallest
in the tropics, tropical agriculture is least able to adapt to even small
changes fast enough to avoid disruption in food supplies.[24]

[22] For rich development of the fundamental issues here, see Sheila Jasanoff's and
Steven Yearley's contributions to this book.
[23] Wilfred Beckerman, "Global Warming and International Action: An Economic Perspective," in *The International Politics of the Environment*, pp. 253–89 (cited at note
6). Also see, in the same collection, Richard N. Cooper, "United States Policy towards
the Global Environment," especially "Choice of a Discount Rate," pp. 300–304. On the
narrow issue of setting a discount rate, contrast William R. Cline, "Time Discounting,"
in *Global Warming: The Economic Stakes*, Policy Analyses in International Economics,
no. 36 (Washington, D.C.: Institute for International Economics, 1992), pp. 72–75. For
a brilliant general ethical critique of discounting as a technique, see Derek Parfit, "Energy Policy and the Further Future: The Social Discount Rate," in *Energy and the Future*, ed. Douglas MacLean and Peter G. Brown, Maryland Studies in Public Philosophy
(Lanham, Md.: Rowman & Littlefield, 1983), pp. 31–37.
[24] On the all-important danger to agriculture and food, especially in the tropics, see
Cynthia Rosenzweig and Martin L. Parry, "Potential Impacts of Climate Change on
World Food Supply: A Summary of a Recent International Study," in *Agricultural Di-*

So far I have said nothing about option 4, the option of aggressively pursuing non-fossil energy sources. Economic activity is a problem, as far as global warming is concerned, only because it injects greenhouse gases—above all, CO_2—into the atmosphere. *This* problem, the carbon emissions from economic activity, would be solved by a switch to photovoltaic, solar/thermal, wind, nuclear fusion, or any other non-carbon-based energy source. I do not think it is technological romanticism to think that a carbon tax invested in R&D on non-carbon energy could simultaneously raise the price of coal, oil, and gas and lower the price of their competitors through technological improvements until some non-carbons would be competitive with the carbons. If we can have the economic activity without the greenhouse gases, there is no problem (as far as global warming goes). Why choose option 3, reductions in economic activity including reductions in employment for an enlarging population, when one could choose option 4, elimination of our carbon addiction?

The primary obstacles appear to be political, not technological. The world is currently run, it seems to me, to a considerable extent to satisfy the preferences of Saudi Arabia, Kuwait, BP, Exxon, the automobile industry, and related interests.[25] The Saudi and Kuwaiti "royal" families/governments were major obstructionists in Rio, and the chair of the Intergovernmental Panel on Climate Change, Dr. Bert Bolin (Sweden), eventually admitted publicly that Saudi Arabia and Kuwait continued to place political obstacles in the way of the IPCC's research;[26] U.S. fossil fuel interests orchestrated a brilliant campaign of disinformation to defeat Clinton's modest 1993 BTU-tax proposal and

---

mensions of Global Climate Change, ed. Harry Kaiser and Thomas E. Drennen (Delray Beach, Fla.: St. Lucie Press, 1993), pp. 87–116, and, for fuller detail, C. Rosenzweig, et al., *Climate Change and World Food Supply* (Oxford: Environmental Change Unit, 1993).

[25] Obviously this is intentionally a great oversimplification. The classic political treatment is Robert Engler, *The Politics of Oil: A Study of Private Power and Democratic Directions* (Chicago: University of Chicago Press, 1961; Phoenix edition, 1967). A more recent economic study, concluding with (generally ignored) recommendations for public policy, is Duane Chapman, *Energy Resources and Energy Corporations* (Ithaca, N.Y.: Cornell University Press, 1983). A Pulitzer-prize-winning history of the modern world centered upon oil is Daniel Yergin, *The Prize: The Epic Quest for Oil, Money, and Power* (New York: Simon and Schuster, 1991).

[26] "IPCC Chairman Says 'Political Wrangling' Now Gives Way to Science and Technology," *International Environment Reporter: Current Reports* 16 (1993): 544. Saudi Arabia and Kuwait—and OPEC more generally—have been the exceptions to a trend of cooperativeness. At the first Conference of the Parties to the Framework Convention on Climate Change (Berlin 1995), the OPEC countries were effectively ejected from the G77 developing-country group "as having maintained an immoral and impractical position too obdurately for the rest of the developing world to bear." See Grubb, "Seeking Fair Weather," pp. 481–82.

nearly defeated what became the pitifully small ($0.043 per gallon) gasoline tax. Are entrenched interests so deeply rooted that serious pursuit even of option 4 will not happen (soon enough)? If so, we would be left with option 3, voluntarily choosing less—perhaps even, no—economic growth.

Yet there are powerful reasons to doubt that option 3 will in fact be chosen. Chris Reus-Smit, in Chapter 5, has marshaled one set of them, which cuts very deep. What justification the modern state has for its pretensions to sovereignty rest on its claims to achieve its "moral purpose," which in this international system (unlike others) is economic growth. To ask states whose raison d'être is the promotion of economic growth to adopt a policy of constraining economic growth might be, in effect, to invite them to commit suicide by undercutting their own rationale, something states rarely do on purpose. This is one fundamental reason to doubt that option 3 is going to be adopted.

Because option 4 seems not to challenge the underlying ideology of the contemporary state but only (!) the preferences that support the current energy regime, the interests arrayed against it are narrower, although obviously very powerful. That it might be adopted is, however, the only glimmer of light I can see in an otherwise possibly dark future. Preferences that block option 4 should, then, I think, be challenged. This is not as radical, Reus-Smit correctly observes, as a challenge to the interests and ideology blocking option 3 would be. There is also a case for mounting that challenge as well, but I hope that aggressive action to moderate global warming does not need to wait until a challenge to the justificatory ideology of the modern state could succeed. This hope rests, of course, on the possibility that option 4, major moves toward alternative energy sources, can be made viable very soon.

Whose preferences, in the end, should count? The "preference"— more accurately, the need—of every human being to engage in enough economic activity to sustain a decent standard of living. Whose preferences should not count? The preferences of those who wish to see the current energy regime, centered around oil, continue undisturbed.

Have I, then, in making option 4 the place to start, settled for a dubious technological fix? Option 4 would do absolutely nothing about, for instance, background international injustices. Option 4 speaks not at all to the many possible reasons other than carbon emissions for fundamental economic change; it would not lead to general economic restructuring. And it is limited in many other respects.

It is crucial, however, to recall the question to which option 4 is the answer: question 4b, what is a fair allocation of emissions of greenhouse gases during the transition to a fair long-term allocation? To this question emerged the answer: in order to avoid violating a minimum standard of fairness, rich countries must begin promptly to reduce progressively their emissions of greenhouse gas rapidly enough that a minimum level of emissions is always available for every human being under the maximum global total of emissions compatible with avoiding climate change to which the poorest could not adjust. Of the two ways to reduce rich country emissions, options 3 and 4, the less painful for everyone without a vested interest in the current energy regime is option 4, to create alternatives to the current energy technologies that spew greenhouse gases into the atmosphere. Immediately.

I have no quarrel with a number of more radical proposals that solve additional problems. But for a start I would be very pleased to see this method of mitigating global warming implemented now.

2

Concepts of Community and Social Justice

WILL KYMLICKA

IF we are to tackle global environmental issues effectively we will need a theory of distributive justice in international relations.[1] The development of such a theory involves a break with traditional Western political theory, which has generally viewed international relations as a Hobbesian "state of nature." Norms of justice are seen as inapplicable in such a state, because there is no mechanism or institution (such as a sovereign authority) to ensure that moral actions are reciprocated. Several commentators, however, have shown that this position cannot be sustained without invoking a more global moral skepticism, one that would equally apply to domestic justice.[2]

Any plausible conception of international justice must include a number of different elements. For example, it will have to reconcile the sometimes competing demands of economic development and environmental preservation. On the one hand, countries in the Third World claim a right of development, including the transfer of re-

For helpful comments and suggestions regarding an earlier draft, I thank Jean Daudelin, John Russell, Susan Donaldson, and Laura Purdy.

[1] Robert Goodin, "International Ethics and the Environmental Crisis," *Ethics and International Affairs* 4 (1990): 91–105.

[2] See, for example, Charles Beitz, *Political Theory and International Relations* (Princeton: Princeton University Press, 1979), part 1, and Marshall Cohen, "Moral Skepticism and International Relations," in *International Ethics*, ed. Charles Beitz (Princeton: Princeton University Press, 1985). It may be true that certain norms of justice require international coordinating mechanisms that do not yet exist. Still, the absence of such mechanisms is not a reason to ignore justice—rather, it gives rise to a duty to create the necessary institutions. Henry Shue, "Mediating Duties," *Ethics* 98, no. 4 (1988): 687–704.

sources and technology to enable them to industrialize and thereby bring their standard of living closer to those in the First World.[3] On the other hand, if development in the Third World continues along its current path, it will lead to the destruction of the global environment. If the Third World ever reaches levels of production matching the First World—with its corresponding use of fossil fuels, CFCs, water, forests, and so on—the world as a whole will become uninhabitable.[4] A theory of international justice must recognize both the legitimate demands for economic development and the need to protect the environment.[5]

Here I discuss another complication that a theory of distributive justice must address—namely, that justice between states does not guarantee justice for substate communities, such as indigenous peoples. I don't just mean the obvious point that elites in Third World countries may impose development/preservation policies in a discriminatory or corrupt way. My concern, rather, is that even if there is a good-faith effort to distribute the costs and benefits fairly, this may still have devastating effects on indigenous peoples, effects that may lead us to question our existing understanding of justice.

The situation of indigenous peoples raises a number of important questions about the presuppositions of both domestic and international justice—for example, the relationship between the claims of individuals, communities, and states; the nature of sovereignty; and the accommodation of cultural differences. I want to look at these questions, and how they affect discussions of social justice and environmental protection.

DISTRIBUTIVE JUSTICE AND INDIGENOUS PEOPLES

Consider the following cases.[6] The government of Bangladesh has encouraged urban poor from the overpopulated heartland to settle and develop the Chittagong Hill Tracts, traditionally occupied by various

[3] For a survey of issues raised by the right to development, see René-Jean Dupuy, *The Right to Development at the International Level* (Alphen aan den Rijn: Sijthoff & Noordhoff, 1980).

[4] Andrew Dobson, *The Green Reader* (San Francisco: Mercury House, 1991), pp. 64–70.

[5] Henry Shue discusses these issues comprehensively in Chapter 1 herein. See also his "Unavoidability of Justice," in *International Politics of the Environment*, ed. Andrew Hurrell and Benedict Kingsbury (Oxford: Oxford University Press, 1992), and "Subsistence Emissions and Luxury Emissions," *Law and Policy* 15, no. 1 (1993): 39–59.

[6] My description of these cases, and of the issues they raise, is heavily indebted to Peter Penz's excellent article on development policy and indigenous peoples. See Penz, "Development Refugees and Distributive Justice: Indigenous Peoples, Land, and the Developmentalist State," *Public Affairs Quarterly* 6, no. 1 (1992): 105–31.

Tibeto-Burman tribes.[7] Similarly, the government of Indonesia has encouraged some of its Javanese citizens to develop and settle western New Guinea, the traditional homeland of the Papuans. In both cases, the state has sponsored the settlement of less populated frontier lands held by indigenous peoples as a response to poverty and landlessness in the heartland.

In both cases, the indigenous peoples are slowly being overrun. In both cases, they have formed resistance movements to secure their claims to govern their traditional homelands and to exclude others from their lands. The government of Brazil also attempted to encourage landless people to settle in Amazonia, but partly retracted the policy under international pressure.[8] In each case, the government justified the settlement policy on the grounds that the lands held by indigenous peoples belonged to the country as a whole, and should be used for the benefit of all, particularly the poorest, both indigenous and nonindigenous. A more intensive population and cultivation of frontier land would promote a more equitable distribution of resources and ensure a better life for more people.

In some cases, the reason why these homelands are relatively less populated is that the indigenous population has been decimated by deliberate killing, and/or by diseases brought by settlers.[9] Obviously, no plausible conception of justice would allow settlers to decimate the local population and then justify their settlement by the fact that there are few indigenous inhabitants left. But the extent of depopulation varies from case to case, as does the culpability of those who are now being encouraged to settle in indigenous homelands. In at least some cases, desperately poor people from the heartland, who may themselves be migrants from other areas, cannot plausibly be said to be responsible for, or the beneficiaries of, acts of genocide committed against indigenous peoples.

For those of us who believe both in resource egalitarianism and in the rights of indigenous peoples, this justice-based argument for settling indigenous lands creates an awkward dilemma.[10] Resource egali-

---

[7] For a detailed discussion of this policy, see Aftab Ahmed, "Ethnicity and Insurgency in the Chittagong Hill Tracts Region: A Study of the Crisis of Political Integration in Bangladesh," *Journal of Commonwealth and Comparative Politics* 31, no. 3 (1993): 32–66.

[8] Andrew Hurrell, "Brazil and the International Politics of Amazonian Deforestation," in Hurrell and Kingsbury, *International Politics of the Environment*, pp. 397–429.

[9] John Hemming, *Amazon Frontier: The Defeat of the Brazilian Indians* (Cambridge: Harvard University Press, 1987). It is estimated that two to five million Indians lived in Brazil when European colonizers first arrived. They now number only around 200,000.

[10] By "resource egalitarianism," I mean the view that justice requires some sort of equality in the distribution of resources. I describe and defend one version of equality of

tarianism insists that there are some limits on the size of the resources that any group can claim—limits to the size of the benefits they can demand or withhold from others. Are the rights demanded by indigenous peoples therefore inconsistent with an egalitarian view of social justice?

There are many reasons to be skeptical of this justification for settlement. Most indigenous groups are not resource-rich—on the contrary, they have already been dispossessed of their most valuable land, and settlement policies take advantage of their vulnerability to appropriate what little is left. Moreover, insofar as indigenous groups do seem better off than the urban poor, this is probably because the poor have outstanding claims of justice against the elites in their own society, or against the First World. Indeed, Brazilian elites use the policy of promoting settlement in the Amazon precisely to deflect efforts at reforming one of the most unequal systems of land ownership in the world.[11] If these obligations of justice toward the urban poor were met, it seems likely that indigenous peoples would no longer have more per capita resources than people in the heartland. Indeed, it may turn out that resource egalitarianism would require redistribution from the heartland to indigenous peoples. It seems reasonable that the obligations of heartland elites and First World countries should be met first, before looking to indigenous resources, if only because there is a much lower level of contact and interdependence between indigenous peoples and the heartland poor, and so the former are less implicated in the latter's condition. So the likelihood that settling indigenous lands would promote resource egalitarianism in the real world is small.

Moreover, experience has taught us that when elites justify settlement policies on the ground that these policies aid the heartland poor, this is often a dishonest rationalization for their own enrichment, and they make sure that they acquire title to the most valuable land or mineral resources.[12] To be sure, this varies from country to country. According to Norman Myers, subsistence farmers (what he calls shifted cultivators) account for about half of the deforestation worldwide, with the other half resulting from a combination of commercial logging, ranching, mining, or infrastructure development (roads,

resources in *Contemporary Political Philosophy* (Oxford: Oxford University Press, 1990), chap. 2.

[11] J. Ronald Engel, "Ecology and Social Justice: The Search for a Public Environmental Ethic," in *Issues of Justice: Social Sources and Religious Meanings*, ed. Roger Hatch and Warren Copeland (Macon: Mercer University Press, 1988), pp. 252–53.

[12] For the extensive role of commercial interests in tropical deforestation, see James Swaney and Paulette Olson, "The Economics of Diversity: Lives and Lifestyles," *Journal of Economic Issues* 26, no. 2 (1992): 1–25.

dams, and so on). So the pressure for settlement is at least in part a demand of the poor, not just a scheme of the elites, and their needs must be addressed if the pressure is to be lessened.[13]

But even when well intentioned, these forms of settlement are almost always unsustainable. Turning rain forest into farms just doesn't work—land clearance leads to soil damage, erosion, and pollution; after one or two harvests the soil is exhausted and the settlers abandon the land and clear more forest.[14] Perhaps the only sustainable forms of use are those already practiced by the indigenous peoples, which is not surprising since they know most about the possibilities and limits of their environment.

Settlement plans are almost always flawed in one or more of these three ways—that is, they serve the rich rather than the poor; and/or they lead to environmental destruction rather than sustainable development; and/or they target indigenous groups that are in fact resource-poor rather than resource-rich. These considerations provide conclusive grounds to reject virtually all existing or proposed settlement policies.

This issue is worth pursuing further, for two reasons. First, people in the Third World may see First World dismissal of settlement policies as hypocritical.[15] After all, much of our wealth came precisely from the settlement of frontier lands that were the traditional homelands of indigenous peoples. This is true of Canada, the United States, Australia, New Zealand, and other developed countries. And it was sustainable development. Even if our current agricultural and mining practices are not sustainable, the fact is that settling and cultivating our frontier has improved the standard of living of millions of people for over a century. And some of the people who benefited from this settlement were urban, landless poor. Indeed, this is the very essence of our national mythologies—people with little to their name could "go West" and start new lives. This may be partly a myth—and certainly the elites also gained from opening up the West—but it is a powerful one.

Moreover, we do not act as if this settlement of the frontier was

[13] Norman Myers, "The Anatomy of Environmental Action: The Case of Tropical Deforestation," in Hurrell and Kingsbury, *International Politics of the Environment,* pp. 432, 447; see also Chapter 4 herein.

[14] See Juan de Onis, *The Green Cathedral: Sustainable Development of Amazonia* (New York: Oxford University Press, 1992). De Onis himself argues that a sustainable form of settlement and development is possible, if a suitable system of land-use zoning is established, under which large areas of rain forest and Indian lands would be preserved intact.

[15] Hurrell, "Brazil," pp. 406–8; Rachel McCleary, "The International Community's Claim to Rights in Amazonia," *Political Studies* 32, no. 4 (1991): 692.

inherently unjust. Of course, there are many aspects of the settlement we now see were unjust—too much coercion and violence was used, too many treaties were broken, and too little land was left for the original indigenous inhabitants. Most people favor some form of compensation for these historical injustices. But this compensation does not take the form of restoring all of the original homelands to their original owners. That is, we act as if the *manner* of settlement was unjust, not the settlement itself, that it was not inherently wrong to expect the indigenous populations in western North America to share at least some of their land with the expanding population in the East.[16]

So the First World should not be too eager to tell governments in the Third World that settling frontier lands is not a viable or just route to development. Of course, the prairies are not the rain forest. And, as I noted earlier, the evidence suggests that few if any settlement policies are enacted out of a sincere commitment to justice. But we should be sensitive to the perception among Third World leaders that we are imposing a double-standard in this area; we should explain our opposition to these policies carefully, rather than dismiss them out of hand. The charge of hypocrisy will be raised, and rightly so, whenever our attitude toward conservation in the Third World appears to be merely self-serving, and so we must address the issue of justice head on.

Another reason for pursuing the issue of justice is that there are important philosophical questions here. While resource egalitarians and indigenous-rights advocates both oppose settlement policies, they do so for different and perhaps conflicting reasons. Many indigenous groups claim that they have an inherent or morally fundamental right to their traditional homelands, including rights to the mineral resources in them, no matter how large these resources are. Brazilian Indians, for example, constitute 0.16 percent of the population, but have rights to 8.5 percent of the land.[17] This seems to conflict with resource egalitarianism, which insists that there are limits on the size of the resources that any group can claim—limits to the size of the benefits they can demand or withhold from others. Resource egalitarianism may support indigenous land claims, but only in a more qualified and conditional form, not as inherent rights.

Moreover, the claim by indigenous peoples that they have inherent rights over their land underlies their stance, not only on settlement

---

[16] For a discussion of these issues, see Jeremy Waldron, "Superseding Historical Injustice," *Ethics* 103, no. 1 (1992): 4–28.

[17] Manuela Carneiro da Cunha, "Custom Is Not a Thing, It Is a Path: Reflections on the Brazilian Indian Case," in *Human Rights in Cross-Cultural Perspective*, ed. Abdullah Ahmed An-Na'im (Philadelphia: University of Pennsylvania Press, 1992), p. 282.

policies, but also on a wide range of other issues, such as self-government rights, treaty rights, hunting and fishing rights, and exemptions
from some forms of taxation. There is a growing international movement toward the international recognition and protection of these
rights, through a separate charter of indigenous rights.[18] The idea that
indigenous peoples have a fundamental moral claim to these rights is
widely seen as inconsistent with liberal egalitarianism, which insists on
equality not only in resources, but also in political rights and legal
status. There seems to be an underlying tension between liberal egalitarianism and indigenous rights that is worth examining, since it arises
in many areas.

In particular, this apparent conflict raises the question whether resource egalitarianism, and liberal justice more generally, is missing
something. Many people will feel uneasy with the idea that justice
could, even in principle, endorse settlement policies that encroach on
the homelands of indigenous peoples. This suggests that liberal justice
is unable to take proper account of the legitimate values and claims of
communities and cultures. Is there a more sensitive way to interpret or
apply principles of justice?

THE ENVIRONMENT AND SOCIAL JUSTICE

There are a variety of objections one could raise to resource egalitarianism. One environmentalist objection says that the whole framework
of distributive justice is inappropriate when discussing land and the
natural environment generally. According to some environmentalists,
the language of distributive justice promotes a view of the natural
world as a "resource" to be used for human consumption. As Robyn
Eckersley puts it, to circumscribe the problem in terms of distributive
justice can "serve to reinforce rather than challenge the prevailing
view that the environment is simply a human resource (albeit a resource to be utilized more efficiently and equitably)."[19] The language
of justice, on this view, reflects and promotes consumerism, or the
"politics of getting." We need to shift from this focus on consuming

---

[18] Benedict Kingsbury, "'Indigenous Peoples' as an International Legal Concept," in
Indigenous Peoples of Asia, ed. R. H. Barnes (Ann Arbor, Mich.: Association of Asian
Studies, 1995), pp. 13–34; Erica-Irene Daes, "Consideration on the Right of Indigenous
Peoples to Self-Determination," *Transnational Law and Contemporary Problems* 3
(1993): 1–11; James Anaya, *Indigenous Peoples in International Law* (New York: Ox-
ford University Press, 1996).

[19] Robyn Eckersley, *Environmentalism and Political Theory* (Albany: State University
of New York Press, 1992), p. 9.

resources to a perspective that distinguishes needs from wants, that emphasizes stewardship rather than possession of the environment.[20]

There are two ways to interpret this objection. Either the language of justice is inherently instrumentalist and consumerist, and hence plays no legitimate role in decisions about the environment; or the language of justice serves a valid but limited function, but needs to be supplemented and constrained by other "ecocentric" considerations. According to the first interpretation, the language of justice is not appropriate for describing the human claims on and human interests in the environment; according to the second, human claims are not the only ones that need to be considered.

I disagree with the first interpretation. I do not believe that the language of justice is intrinsically linked to a conception of human relationships as instrumental or consumerist.[21] But even if we replace the language of justice with another language that emphasizes our connectedness with nature and with each other, we will not have solved the problem at hand. For the question facing us is precisely which groups of people should be "connected" with which tracts of land. The task of assigning territory to people remains, whether we ask it in the language of "rights to resources" or in the language of "responsibilities over the natural environment." And a plausible principle to guide such decisions is resource egalitarianism—namely, that decisions about territory should be made in such a way that everyone has the same opportunity to benefit.

Many real-world settlement policies reflect greed and consumerism. But if these policies are genuinely motivated by equality, then their aim is to help meet the basic needs of the poor rather than merely enriching the elites.[22] In such cases, we cannot avoid the question of fairly allocating people and land.

The second interpretation says that although the language of justice is appropriate for describing and adjudicating the claims of human beings, other living beings or ecosystems also have moral standing and moral claims that must be weighed against the claims of human jus-

---

[20] Ibid., p. 18; Andrew Dobson, *Green Political Thought* (London: Unwin Hyman, 1990), pp. 91–92. More generally, some people see a conflict between the values of social justice and ecology, each of which must be compromised in the name of the other. To insist on perfect equality or absolute social justice, they say, would amount to demanding "fair shares in extinction." Sandy Irvine and Alec Ponton, quoted in *Green Political Thought*, p. 173.

[21] See my *Liberalism, Community, and Culture* (Oxford: Oxford University Press, 1989), pp. 122–26, and *Contemporary Political Philosophy*, pp. 164–69.

[22] It is important to note that two-thirds of Brazil's population is living at the subsistence level. McCleary, "Rights in Amazonia," p. 700.

tice. We must adopt a nonanthropocentric moral theory that recognizes not only the rights of humans (for example, to a fair share of resources), but also the rights of animals or nature. And once we do this, our obligations to nature override any settlement policy, even one otherwise required by justice to humans.

For example, Eric Katz and Lauren Oechsli have recently argued that if we remain within the language of justice, then Brazil should be free to develop the Amazon. It is unjust to preserve the Amazon for the benefit of a few Indians, or for the benefit of the developed world, when it can dramatically improve the well-being of many poor Brazilians. As they put it, "The demand for anthropocentric justice dooms the preservation of the natural environment."[23] But, they go on to say, the preeminence of justice in the debate reflects the fact that "the policy discussion has been limited to a consideration of human interests."[24] If we shift to a nonanthropocentric worldview, then issues of justice become secondary. They suggest that the debate about who should control the rain forest is like a debate between two criminals over how to distribute the gains from robbing and murdering an innocent bystander. Since the bystander's rights to person and property were violated, neither criminal has any right to those gains. Similarly, no one has any right to the benefits from a settlement policy that violates the inherent rights of the rain forest. As they put it, "Questions of the trade-off and comparisons of human benefits and questions of justice for specific human populations do not dominate the discussion. . . . The competing claims become insignificant in light of the obligations owed to [the rain forest] . . . the obligation to the rain forest makes many of the issues about trade-offs of human goods irrelevant.[25] Although there may seem to be a dilemma if we focus solely on fairness between groups of humans, "once we move beyond the confines of human-based instrumental goods, the environmentalist position is thereby justified, and no policy dilemma is created."[26]

I call this the ecocentric argument against settlement. I can't evaluate it in depth, since it raises questions of the intrinsic value of the environment, and how these nonhuman claims are to be weighed

[23] Eric Katz and Lauren Oechsli, "Moving beyond Anthropocentrism: Environmental Ethics, Development, and the Amazon," *Environmental Ethics* 15 (1993): 56. They mention the possibility that Brazil could be compensated by the First World for the benefits it forgoes by preserving Amazonia, but think this unlikely, and so are looking for an argument for preserving the Amazon that would "trump" considerations of human justice.

[24] Ibid., p. 56.
[25] Ibid., pp. 57–58.
[26] Ibid., p. 58.

against the claims of people to the satisfaction of their basic needs.[27] I will just make two comments. First, even if we think that certain parts of the Third World should be preserved from development for nonanthropocentric reasons, this does not render issues of justice irrelevant. It would still be essential to ensure that the costs of environmental preservation were distributed fairly and did not fall disproportionately on the Third World poor.

Second, it would be misleading to view the ecocentric argument against settlement as a defense of the claims of indigenous peoples. For one thing, we can't assume that the lands held by indigenous peoples are always the most valuable in terms of the claims of nonhuman animals or ecosystems. The traditional homelands of indigenous peoples are often seen as "wilderness" areas,[28] and wilderness areas generally contain a great wealth of nonhuman life that would be harmed by settlement. But we can't assume that the land identified by an ecocentric theory as particularly valuable is going to be identical to the land identified by indigenous peoples as their own. It is important to note that more intensive development of the heartland also harms animals and ecosystems, and so an ecocentric theory may find that that option also violates the claims of nonhuman life.[29]

More important, the ecocentric argument against settlement policies also argues against attempts by the indigenous peoples themselves to develop their resources. If the basis for preserving indigenous land from settlement is that it is wilderness; and if, as Daly and Cobb argue, the only pattern of human habitation and use compatible with wilderness is hunting and gathering,[30] then the ecocentric argument would preclude indigenous people from shifting toward a more agricultural or urbanized lifestyle. It would preclude any attempt at modernization undertaken within their traditional homelands.

[27] Wendy Donner discusses these issues in Chapter 3.
[28] As Eckersley notes, there is an element of ethnocentrism in many everyday references to "wilderness," since they ignore the fact that indigenous peoples have often practiced ecological management of these "wilderness" areas (for example, by burning out the undergrowth). Eckersley, *Environmentalism*, p. 40. Still, although these areas cannot be seen as "pristine" wilderness, they have not been subject to the radical transformations entailed by either urbanization or agricultural use.
[29] Indeed, one of the reasons why Brazil objects to the international preoccupation with the Amazon is that it ignores other environmental problems in Brazil, particularly the enormous environmental degradation around Brazil's urban centers. Hurrell, "Brazil," p. 421.
[30] Herman Daly and John Cobb Jr., *For the Common Good: Redirecting the Economy toward Community, the Environment, and a Sustainable Future* (Boston: Beacon Press, 1989), p. 253.

Paradoxically, then, the ecocentric argument would reinforce the efforts of the Brazilian government to deny the rights of Indians. As Manuela Carneiro da Cunha notes, the Brazilian government has tried to reinterpret Indian land rights so that they only apply to "real Indians"—namely, those who have maintained their "traditional culture." The intended result is that "ultimately, there would be virtually no holders of Indian rights and coveted lands would become available."[31] As da Cunha notes, the ecocentric argument denies the dynamic nature of ethnic identity. Thus it would have the same result of limiting Indian claims to groups whose cultural practices and ethnic identity have become frozen in time.

Many indigenous peoples have chosen not to adopt a more modern way of life. They do not want modernization forced upon them, but they do not want to be prevented from modifying their traditional lifestyles either. Many indigenous groups have in fact moved some way toward a more urban and agricultural lifestyle, but they demand the right to decide for themselves what aspects of the outside world they will incorporate into their cultures, and the right to use their traditional resources in that process. The ecocentric argument, therefore, is not really compatible with the claims of many indigenous peoples.[32]

COMMUNITY AND SOCIAL JUSTICE

Another set of objections to resource egalitarianism focuses on its alleged "individualism" and the need to adopt a more "communitarian" approach to justice. But what does it mean to incorporate community in a theory of justice? It is a central tenet of Green theorists that we must situate individuals as members of communities. The basic unit of political theory, they say, should be the "individual-in-community," not the atomistic individual of liberalism, which abstracts the individual from social relationships.[33]

There is some truth to the claim that resource egalitarianism neglects the way people are situated in communities. There are many versions of this claim, each of which has different implications for social justice. When Green theorists talk about the "individual-in-com-

[31] Da Cunha, "Custom Is Not a Thing," p. 284.
[32] As da Cunha notes, many environmentalist discussions of Brazil have seen indigenous peoples as "part of the natural scenery." There has been a "naturalization" of indigenous groups, who are not seen as "agents with their own specific projects." Ibid., pp. 286–87.
[33] Daly and Cobb, For the Common Good, pp. 159–75; Eckersley, Environmentalism, pp. 53–55.

munity," the sort of community they usually have in mind is some form of subnational, geographically defined community. This may be an existing territorial subunit, such as a province or county, or a new, environmentally defined unit, such as a "bioregion."[34] Decisions made at the community level are seen as more environmentally responsible than decisions made society wide, since people are more aware of their local environment and are tied to each other in bonds of both economic and ecological interdependence. A substantial part of recent Green theorizing has focused on how to decentralize power to such communities, and how to promote people's sense of identity with them. Proposals range from a more decentralized form of federalism, to greater democratization at the local level, to the communal ownership of some or all land. I call this the decentralist argument.[35]

There is much to be said in favor of these proposals. Certain kinds of community, and collective action, are only possible in smaller groups. Membership in these groups can give a sense of belonging and participation. For that reason, we should protect these smaller groups from being undermined by economic or political pressure from the larger society—for example, by reducing the constant pressure in a capitalist society for people to migrate from one region to another for economic reasons, thereby undermining the sense of local community.

It is important not to exaggerate the point. There are many aspects of economic, social, and environmental policy that can only be effectively dealt with at the federal level. Too much decentralization of power may result, not in the empowering of smaller communities, but simply in leaving everyone powerless in the face of global economic and political trends.

In any event, this decentralist argument does not necessarily help us reconcile justice and the claims of indigenous peoples. It may justify a general devolution of power from larger to smaller jurisdictions or communities. It may also justify the provision of resources in the form of communal property rather than individual property. But it can't explain why these powers and resources are distributed differentially among smaller communities, as is implicit in the claims of indigenous peoples. Nor can it explain why these local communities should be defined on the basis of ethnic criteria.

[34] See, for example, Dobson, *Green Political Thought*, pp. 117–22, and *Green Reader*, pp. 73–83, and Eckersley, *Environmentalism*, pp. 160–76.
[35] For discussions of decentralization as a strategy for empowering citizens and protecting the environment, see the chapters by Smitu Kothari and Joseph Camilleri in this volume.

For example, the decentralist argument might justify devolving powers from the federal to the state level in Brazil, or to the bioregion of Amazonia. But this wouldn't help the indigenous peoples, since settlers constitute the overwhelming majority both at the state level and in Amazonia as a whole. And indeed, the governors of the states in Brazil that include the Amazonian Indians are in favor of greater settlement and development and have bitterly opposed the plans of the federal government to create large native reserves.[36]

This has created a paradox in government policy toward the Amazon. The Brazilian federal government has enacted some policies to protect the environment and indigenous communities. But the closer one gets to the proposed development site, the more likely it is that the majority of the local population opposes these policies and favors development. Conversely, the greatest support for these policies often comes from the international community. This "democratic deficit" has made it virtually impossible for the federal government to actually enforce its policies.[37] Decentralization would make it even more difficult to ensure respect for indigenous rights.

What indigenous peoples demand is not a general decentralization, but rather that political boundaries be redrawn, based on ethnic criteria, to give them a self-governing enclave. The idea of protecting local communities cannot by itself explain this aspect of indigenous rights.

Moreover, the decentralist argument doesn't tell us why per capita resources should differ between communities, be they states, counties, or bioregions. Even if we accept that resources should be distributed in the form of communal property, we still need to know why indigenous communities should have more property on a per-capita basis than nonindigenous villages, rather than distributing property on an equal per-capita basis.

What the decentralist argument seems to be missing is the importance of certain culturally defined groupings. Certain cultural groups—including indigenous cultures, and many other ethnocultural groups as well—claim that they have special rights to both powers and resources. If we are to assess this claim, it is important to see people not only as "individuals-in-community," but also as "individuals-in-cultures."

Why does it matter that indigenous peoples are distinct cultures? There are at least three answers to this question, which I call the cul-

[36] De Onis, *Green Cathedral*, p. 227.
[37] See Jean Daudelin, "L'environment comme Cheval de Troie? Le Cas de l'Amazonie brésilienne" (paper presented to the annual meeting of the Canadian Political Science Association, Ottawa, June 1993).

tural relativism, minority disadvantage, and national self-determination arguments.

Cultural Relativism

Cultural relativism is the view that each culture has its own standards of justice and morality, which must be accepted as valid for it, since there are no rational grounds on which to prefer one culture's views to another. Hence we should not interfere with another culture on the basis of some allegedly universal theory of justice.

This is a common theme in recent communitarian critiques of liberalism. Many communitarians claim that liberals misinterpret justice as an ahistorical and external criterion for criticizing the ways of life of every society. Utilitarians, liberal egalitarians, and libertarians may disagree about the content of justice, but they all seem to think that their preferred theory embodies a universally applicable standard. They do not see it as a decisive objection that their theory may be in conflict with local beliefs.

Michael Walzer argues that this quest for a universal theory of justice is misguided. There is no such thing as a perspective external to the community, no way to step outside our history and culture. The only way to identify the requirements of justice, he claims, is to see how each particular community understands the value of social goods. A society is just if it acts in accordance with the shared understandings of its members, as embodied in its characteristic practices and institutions. Hence identifying principles of justice is more a matter of cultural interpretation than of philosophical argument.[38]

Walzer's theory is, of course, a form of cultural relativism, and it is beyond the scope of my discussion to enter that age-old philosophical debate. But it is worth noting two common objections to communitarian attempts to define justice in terms of a community's shared understandings. First, and paradoxically, Walzer's theory violates one of our deepest shared understandings. According to Walzer, slavery is wrong in our society because our society disapproves of it. But most people believe that we disapprove of slavery because it is wrong. Its wrongness is a reason for, not the product of, our shared understanding. Second, there may not be many shared understandings about justice, especially if we attend not only to the voices of the vocal and powerful, but also to the weak and marginalized. In order to resolve these disagreements, we need to assess competing understandings in the light of a more general conception of justice. So even if we start

[38] Michael Walzer, *Spheres of Justice* (Oxford: Blackwell, 1983).

with local understandings, as Walzer suggests, we are driven by the existence of disagreement, and our own critical reflection, toward a more general and less parochial standpoint.[39]

Yet even if these familiar objections can be answered, cultural relativism is distinctly unhelpful in this context. For one thing, it is not true that endorsing cultural relativism leads to the principle of noninterference. What cultural relativism says is that each culture rightly acts on the basis of its own moral code. So if the indigenous people see the Amazon as their homeland and birthright, then they can rightfully defend their lands. But if Brazilians see the Amazon as their frontier and the source of their future riches, then they can rightfully settle their frontier. This is part of their national identity and mythology, just as settling the frontier is part of our national mythology.

Cultural relativism does not help us decide which side is in the right, since both cultures are acting in accordance with their own understanding of the meaning of the Amazon. Cultural relativism says that both cultures can act on their own morality, which cannot be judged from the outside. So it would protect indigenous peoples from the demands of universalist theories of justice, but not from the demands of the particularist theories held by surrounding cultures.

Also, while cultural relativism might protect indigenous peoples from the redistributive demands of universalist justice, it would equally insulate the First World from any obligation to redistribute resources to the Third World (including to poor indigenous groups in the Third World). If a Western country sees itself as rightfully owning all of its wealth, then we cannot criticize it for refusing to share it with the developing world.

More generally, cultural relativism reduces intercultural relations to issues of mutual advantage, rather than issues of justice. But that is precisely what we are trying to get away from when discussing social justice and global environmental change.

Minority Disadvantage

If we wish to respect cultural difference without falling into the trap of relativism, we need to find some more specific feature of indigenous cultures that might justify special rights. One obvious possibility is that the indigenous peoples are a *minority* culture, and as such have certain disadvantages and vulnerabilities that require special resources.

[39] I discuss this in *Liberalism*, pp. 67–69, 231–33. See also Amy Gutmann, "The Challenge of Multiculturalism to Political Ethics," *Philosophy and Public Affairs* 22, no. 3 (1993): 171–206.

According to this argument, special rights are needed to ensure that all citizens are treated with genuine equality. On this view, "the accommodation of differences is the essence of true equality," and special rights are needed to accommodate our differences.[40] Many liberal individualist critics reject this argument. They argue that a system of universal individual rights already accommodates cultural differences, by allowing each person the freedom to associate with others in the pursuit of shared religious or ethnic practices. Freedom of association enables people from different backgrounds to pursue their distinctive ways of life without interference. Every individual is free to create or join various associations, and to seek new adherents for them, in the "cultural marketplace." On this view, giving political recognition or support to particular cultural practices or associations is unnecessary and unfair. It is unnecessary, because a valuable way of life will have no difficulty attracting adherents. And it is unfair, because it subsidizes some people's choices at the expense of others.[41]

An equality-based argument for special rights, therefore, must show that some groups are unfairly disadvantaged in this cultural marketplace, and that political recognition and support rectifies this disadvantage. I believe this is possible. What the liberal individualist view ignores is that the cultural marketplace may be unfairly biased against certain groups. Minority cultures are often vulnerable to economic, political, and cultural pressure from the larger society. The viability of their communities may be undermined by economic and political decisions made by the majority. They could be outbid or outvoted on resources and policies crucial to their survival as a culture. The members of the majority culture do not face this problem. Community rights, such as land claims, veto powers, language rights, guaranteed representation, and local autonomy, can help rectify this disadvantage by alleviating the vulnerability of minority cultures to majority decisions. The exact nature of these rights will vary with each culture. Indigenous peoples are often the most vulnerable because of their size, the great distance between their cultures and European cultures, and their susceptibility to disease. As a result they often require considerable protection, in the form of reserved lands and self-governing powers.

Hence special rights compensate for unequal circumstances that put the members of minority cultures at a systemic disadvantage in the

[40] As stated by the Supreme Court of Canada, in explaining the meaning of the equality guarantees in the Charter of Rights and Freedoms. See *Andrews v Law Society of British Columbia* (1989) 10 C.H.R.R. D/5729 (S.C.C.).

[41] See, for example, R. Knopff, "Language and Culture in the Canadian Debate," *Canadian Review of Studies in Nationalism* 6, no. 1 (1979): 66–82.

cultural marketplace. I call this the equality argument, and have defended it at length elsewhere.[42] This is similar to the debate over another group-specific policy: affirmative action for women or people with disabilities. Like special rights for minority cultures, affirmative action programs asymmetrically distribute rights or opportunities on the basis of group membership. Proponents argue that they are required for genuine equality. Critics respond that the economic marketplace (like the cultural marketplace) already respects equality, by treating job applicants without regard for their group membership; however, an equality-based argument for group-specific affirmative action can be made if the actual operation of the economic marketplace works to the disadvantage of certain groups. As with special rights for minority cultures, the equality argument for affirmative action seeks to show how the structure of universal individual rights is intended to treat all people equally, but in fact works to the disadvantage of the members of a particular collectivity. Many group-specific claims can be seen in this way—namely, as compensating for the disadvantages and vulnerabilities of certain groups within the structure of universal individual rights. Of course, affirmative action for women or people with disabilities differs in many ways from community rights for minority cultures, since they are compensating for very different kinds of injustices. The former is intended to help disadvantaged groups integrate into society, by breaking down unjust barriers to full integration. The latter is intended to help cultural communities maintain their distinctiveness, by protecting against external pressures to assimilate. This means that the former are (in theory) temporary, whereas the latter are permanent, barring dramatic shifts in population.

This equality argument for special rights and resources is not unlimited. At some point, additional resources assigned to indigenous peoples would not be necessary to protect against vulnerabilities, but rather would simply provide unequal opportunities to them. In these circumstances—which may not exist anywhere on the globe—indigenous peoples would have an obligation to redistribute some of their wealth to other peoples. Even in these circumstances, the exact form of redistribution is important. Given the dependence of indigenous peoples on their land, a radical redistribution could have devastating effects on the sustainability of the culture. Indigenous peoples should be given the time to "progressively economize" on their use of resources and thereby adapt their cultures to the requirements of justice.[43]

[42] Will Kymlicka, *Multicultural Citizenship: A Liberal Theory of Minority Rights* (Oxford: Oxford University Press, 1995), chap. 6; cf. *Liberalism*, chaps. 7–10.
[43] Penz, "Development Refugees," p. 122.

Some proponents of indigenous rights object to the idea that their land claims should be viewed as "special protection rights" that protect a vulnerable minority culture from the majority. They worry that this will promote the view that indigenous peoples should be treated paternalistically as wards, or that land rights should only be granted to indigenous communities that maintain their "traditional" culture. To avoid these dangers, indigenous lands claims should be seen as ordinary, historical property rights based on prior occupancy.[44] From a strategic point of view, this may be the best way to defend indigenous rights in particular circumstances. But, as I discuss below, from the point of view of egalitarian justice, claims of prior occupancy are very weak. Indeed, the whole point of resource egalitarianism is that "first-come, first-serve" is not a valid theory of justice.

National Self-Determination

Indigenous peoples are not just minority cultures, they are also "colonized" minorities. What I mean is that they are distinct cultural communities that were previously self-governing, but whose homeland has now been included in a larger state against their will. They occupied and governed their lands before the state was even in existence.

I think this is very important. The point isn't that indigenous peoples "were here first," and so have property rights as the initial appropriators of the land. There are several reasons why "we were here first" is not enough to justify indigenous land claims. For one thing, it is not clear that the initial appropriation by indigenous peoples was devoid of force or fraud, as is required by theories of justice in initial acquisition. There is strong evidence of conflict between different groups of indigenous peoples, even before European settlement, and the land occupied by a tribe today may well have been acquired by force or fraud from another tribe. If such force against earlier indigenous tribes gives rise to legitimate title, then why can't European settler groups use force against the current indigenous owners? More important, the underlying theory of property is untenable. People should only be able to insist on exclusive use of parts of the natural world if they leave "enough and as good" for others. That condition may have applied when indigenous peoples originally occupied their homelands, but it is no longer true today. Since there is no unclaimed land left for the heartland poor to appropriate for themselves, any claims to property must be judged against a standard of equality. Moreover, accepting a theory of property based on "first-come, first-served" would

---

[44] Da Cunha, "Custom Is Not a Thing," pp. 284–85.

have the same unintended effect as endorsing cultural relativism—that is, while it may insulate indigenous peoples from redistributive demands, it would also insulate wealthy countries from any obligation to redistribute resources to the Third World (including indigenous groups in the Third World).

So the point is not simply that indigenous peoples were the initial appropriators of the land. The point, rather, is to question the boundaries of the political community. This highlights a problematic underlying assumption of the equality argument. The equality argument assumes that the state must treat its citizens with equal respect. But there is the prior question of determining which citizens should be governed by which states. For example, should the Brazilian government have the authority to govern the Yanonami Indians, or are they self-governing? Does the government of Bangladesh have the authority to govern the indigenous peoples in the Chittagong Hill Tracts, or are they rightfully self-governing? After all, they were originally self-governing, and had the balance of power been different, they could have maintained their independence. They only lost their self-government as a result of coercion and colonization. They view this, rightly I think, as a violation of their inherent right to self-government. (For this reason, it is appropriate that settlement policies are often described by the government as "colonization" policies.)

Under international law, all "peoples" are entitled to self-determination—in other words, an independent state. This principle has been applied to grant independence to overseas colonized peoples who were forcibly included in European empires; however, it has not been applied to internal colonized peoples, such as indigenous peoples, who were forcibly included in larger contiguous states. There is no principled reason for this differential treatment of internal and overseas colonized peoples. Indigenous groups in Brazil, as in Africa, are peoples—that is, previously self-governing, territorially concentrated, culturally distinct societies. The process of colonization was just as coercive in Brazil as in Africa, and its effects have been just as devastating.

I don't mean that indigenous peoples should all demand or be granted independent states. This is not a viable or desirable option for all such groups. Nevertheless, their incorporation into a larger state is only legitimate if it is a voluntary act of federation. Agreeing to enter a federation with other cultures is one way in which a people can exercise their right of self-determination. And if we ask what are the terms under which two or more peoples would voluntarily federate, it seems clear that indigenous peoples would only choose to enter such a federation if it recognized their inherent rights of self-government over their traditional homelands.[45] If not independence, they would at least demand self-government and recognition as a distinct people. These demands are at the heart of the recent proposal for an international charter of indigenous rights.

The fact that indigenous peoples can be seen as peoples with inherent rights of self-determination does not absolve them from redistributive obligations.[46] After all, we do not want to absolve the citizens of First World countries of their redistributive obligations to the Third World just because they are independent peoples with rights of national self-determination. On the other hand, it does mean that if an indigenous community has an obligation to progressively economize on the use of resources, this should *not* take the form of an involuntary appropriation of their lands.

Instead, it should probably take the form of a resource tax.[47] It would be up to the indigenous peoples themselves to decide how to manage their resources to pay for this tax. Some communities may decide to sell some of their land, or lease it, or develop some of their mineral wealth, or invite outside people to develop the wealth for them. These are the options facing First World countries in deciding how to fulfill our obligations of justice, and this sort of decision is rightly decided by self-governing peoples.

So there are two ways in which principles of justice should be adapted to accommodate the special status of indigenous peoples— extra resources may be required to rectify the disadvantages they face as minority cultures; and indigenous peoples, as colonized peoples with inherent rights of self-government, should be free to decide for themselves how to manage their traditional homelands in accordance with principles of justice.

These two arguments also apply, with lesser force, to some nonindigenous minority cultures (such as the French in Canada). They too may face disadvantages because of their minority status, and may have rights of self-government that were lost when they were forcibly incorporated into a larger federation. If so, then resource egalitarianism

---

[45] .Kymlicka, *Multicultural Citizenship*, chap. 6. Cf. Kymlicka, "Individual and Community Rights," in *Group Rights*, ed. Judith Baker (Toronto: University of Toronto Press, 1994).

[46] Nor does it absolve them from the obligation to respect the human rights of their own individual members. A community's right to self-determination vis-à-vis other communities does not include or entail the right to oppress people *within* the community. I discuss the relationship between "internal" (intragroup) and "external" (intergroup) aspects of the right of self-determination in *Multicultural Citizenship*, chaps. 3 and 8.

[47] Penz, "Development Refugees," p. 121.

may need to be adapted to provide special rights and resources for some nonindigenous minority cultures as well.

I HAVE just touched on some of the issues about community and culture that need to be addressed in a global theory of distributive justice. There are many other issues, such as how to measure the sorts of disadvantages faced by minority cultures; and how various forms of federalism, decentralization, and secession can accommodate the special needs of minority cultures. We also need to think carefully about how to maintain social unity and stability in a society that recognizes and institutionalizes these sorts of cultural differences. Many countries have resisted the idea that indigenous peoples are "nations" on the ground that this might promote secessionism.[48] Even where the national minority is not secessionist, the fact that citizens see themselves as belonging to distinct nations may affect the functioning of a democratic society. Many people worry that citizens divided by rival national identities may not be willing to make the sort of sacrifices for each other that are needed for a stable and just democratic society.[49]

There are also important questions about how to make use of the special ecological wisdom held by indigenous peoples; and whether sacred sites should be exempted from theories of distributive justice. And there are questions about how these substate communities can be adequately represented in international debates about global justice. Given the special claims that indigenous peoples have to their lands, it is clearly essential that they be properly represented at such debates. To date, this has not happened.[50]

The answers to these questions are often elusive. What is clear is that we must develop an approach to justice that is sensitive to community. Neither mainstream conceptions of social justice nor the more recent environmentalist theories have tackled the many dilemmas raised in this area. We need a theory that requires the First World to help Third World countries develop, but that does so in a way that

[48] Da Cunha, "Custom Is Not a Thing," p. 282.

[49] See, for example, Alan Cairns, "The Fragmentation of Canadian Citizenship," in *Belonging: The Meaning and Future of Canadian Citizenship*, ed. William Kaplan (Montreal: McGill-Queen's Press, 1993), pp. 181–220, and Arthur Schlesinger, *The Disuniting of America* (New York: Norton, 1992). I try to respond to these concerns in *Multicultural Citizenship*, chap. 9, and in "Social Unity in Liberal States," *Social Philosophy and Policy* 13, no. 1 (1996): 105–36.

[50] For a discussion of the political and epistemological obstacles to the representation of such groups, and to the recognition of their knowledge and authority, see Chapters 5, 7, and 8 in this volume.

does not undermine either the environment or indigenous cultures. In short, we need a theory that combines a commitment to international (and intercultural) redistribution, environmental protection, and respect for cultural difference.

3

Inherent Value and Moral Standing in Environmental Change

WENDY DONNER

WHAT obligations do humans have to the environment? Do individuals such as plants and nonhuman animals and wholes such as species, ecosystems, and the entire biotic community have inherent value or moral standing? In this chapter I critically survey some positions on this question that have been prominent in recent debates in environmental ethics.

One fundamental assumption underlying much of the dialogue in the global change and social justice project has been that the framework for analyzing these issues should be human-centered or anthropocentric. For the most part, I agree. Since the work of philosophers is often taken to be the probing questioning of assumptions, however, in this chapter I look at some environmental theories that reject this assumption. Debate on environmental issues in many cases seems to be carried on in two different worlds. In one world are those theorists propounding what Arne Naess calls anthropocentrism or "shallow ecology."[1] Shallow ecologists or human-centered theorists see environmental problems as problems of appropriate resource use. The environment has no moral standing in its own right, and questions of who counts or who has standing are all questions about which human claims, preferences, or rights deserve to be given weight or taken into account in moral deliberations. The other world is inhabited by radical environmentalists who see all such human-centered approaches

[1] Arne Naess, *Ecology, Community, and Lifestyle*, ed. and trans. David Rothenberg (Cambridge: Cambridge University Press, 1989), pp. 14–20.

merely as ethics for managing the environment, rather than as genuine environmental ethics. These theorists regard an appropriate environmental ethic as one centered directly on the environment, not on human ends, purposes, or concerns. The activists who fought the British Columbia government's decision to allow clearcut logging in Clayoquot Sound on Vancouver Island's old-growth forest were not concerned about whether these forests were being used wisely or in a sustainable way—they did not want these forests used at all. They considered these forests valuable in themselves, apart from any human ends or purposes, and wanted them to be left untouched.

One of my aims here is to bridge the gap between these two worlds by giving some of the flavor of the radical environmentalist viewpoint. Theorists who uncritically adopt an anthropocentric approach should make every effort to listen to and attempt to address the legitimate concerns of radical environmentalism. No discussion about human interaction with the environment would be complete without an exploration of some of these more radical theories. Yet when these theories or arguments are mentioned in the dialogue on global environmental change and social justice, it is sometimes done in a cursory way.[2]

The literature of radical environmentalism is vast and complex. I restrict my focus to several theories that attempt to widen the moral community and to view nonhuman aspects of the environment as our kin: the extensionist animal rights/animal welfare theories of Tom Regan and Peter Singer; the land ethic of Aldo Leopold and J. Baird Callicott; deep ecology; and the ecofeminism of Karen Warren, Val Plumwood, and others. Environmental philosophers often draw upon a concept of community as a basic element of an acceptable environmental ethic. The land ethic of Leopold and Callicott, for example, advances the claim that humans ought to expand the notion of the moral community to which we belong to include plants, animals, species, and the biotic community.[3] Put alternatively, the claim is that we ought to conceive of the self as part of nature, as identified with nature, or as not separate from, split off from, or alienated from nature. These new theories are not without their problems.

---

[2] See, for example, Joseph Camilleri, "Global Impoverishment and the National State" (Chapter 6), and Christian Reus-Smit, "The Normative Structure of International Society" (Chapter 5). Will Kymlicka, Iain Wallace, and David Knight (also represented in this book) take these positions more seriously.

[3] Aldo Leopold, *A Sand County Almanac* (New York: Oxford University Press, 1989); J. Baird Callicott, *In Defense of the Land Ethic: Essays in Environmental Philosophy* (Albany: State University of New York Press, 1989), and "The Search for an Environmental Ethic," in *Matters of Life and Death: New Introductory Essays in Moral Philosophy*, 3d ed., ed. Tom Regan (New York: McGraw-Hill, 1993), pp. 323–82.

Moral Extensionism

I begin with the traditional philosophical problem of moral standing. There are certain criteria a being must meet if its interests and well-being are to be taken into account in moral decisions. Work on this problem gained impetus from many human-centered dilemmas, two of the most anguishing being abortion and euthanasia. Any attempt to resolve such questions has to start by coming to grips with what gives humans independent moral status. Most of the proposals and arguments ground independent moral standing in the possession of some degree of consciousness. On one end of the continuum of proposals on the question, the utilitarians Jeremy Bentham and Peter Singer, among others, argue that the mere capacity to feel pleasure and pain is sufficient to confer moral status. In arguing that animals merit such consideration, Bentham claims that "the question is not, Can they reason nor Can they talk? but, Can they suffer?"[4] On this view of things, if a being can feel pleasure or pain, its pain or pleasure matters to it, and thus *it* matters, and we must take account of this in our moral considerations. This argument, deceptively simple and logically compelling, has persuaded many to reject the traditional view that animals lack moral standing.

Other views of moral standing also ground standing in consciousness, but are more demanding. At this other end of the continuum of proposals stands the Kantian, who argues that more highly developed features of consciousness, such as rationality, self-awareness, the ability to communicate with others and use language, the ability to reach contracts and engage in a community, are essential features of persons. But this view faces the problem that although it excludes most or all nonhuman animals, it also excludes a significant number of humans, including infants. This consequence is so repellent that we are forced back toward the other end of the continuum, where standing is grounded in less demanding features of consciousness.

Once standing is grounded in the possession of consciousness, especially sentience, it is but a short step to the realization that logical consistency requires that if humans have standing because they have consciousness, other animal species with consciousness, especially those at the higher end of the evolutionary scale, also deserve stand-

---

[4] Jeremy Bentham, quoted by Martin Benjamin, in "Ethics and Animal Consciousness," in *Social Ethics*, 3d ed., ed. Thomas Mappes and Jane Zembaty (New York: McGraw-Hill, 1987), p. 481. Peter Singer's views are set out in *Animal Liberation*, 2d ed. (New York: Random House, 1990), and *Practical Ethics* (Cambridge: Cambridge University Press, 1979).

ing. Peter Singer uses the term "speciesism" as a parallel to racism and sexism to hurl at those who would deny standing to animals with detectable qualities of consciousness. Tom Regan defends animal rights on the grounds that animals are subjects of a life that from their point of view can be better or worse.[5] Regan says that "for example, if a given theory considers *human* pain and suffering morally relevant, but denies the moral relevance of the pain and suffering of the black bear, then it seems to be rationally defective. For pain is pain, and pain is in itself undesirable, to whomsoever it may occur, whether beast or human."[6] Thus these theorists conclude that sentient animals ought to be included in our moral community and regarded as kin.

Peter Singer, who grounds standing in sentience, also limits it to creatures who have sentience and wants to draw the border there. Our sympathy can be extended to animals. We can understand, according to Singer, what it means to think like a bear, but this imaginative exercise used to enlarge the moral community does not extend beyond conscious creatures. Singer says, "Suppose that we apply the test of imagining living the life of the weed I am about to pull out of my garden. I then have to imagine living a life with no conscious experiences at all. Such a life is a complete blank; I would not in the least regret the shortening of this subjectively barren form of existence. This test suggests, therefore, that the life of a being that has no conscious experiences is of no intrinsic value."[7]

Opponents argue that this amounts to conferring value in too anthropocentric or egocentric a manner—that is, by arguing that something has value if *we* can sympathetically identify with it. Whether such sympathizing is an exercise in egoism or an extension of our moral community remains in contention.

At this juncture we reach the boundary that divides those theories that resolutely stick to consciousness as a basis of standing from those that attempt to leap beyond. Whether or not we attempt to grant value to nonconscious elements of the natural world such as plants, mountains, deserts, entire species, and ecosystems, depends partly upon whether we take this leap from an individualist to an holistic paradigm.

J. Baird Callicott argues that this leap is necessary to get beyond the

---

[5] Tom Regan, *The Case for Animal Rights* (Berkeley: University of California Press, 1983), p. 243.

[6] Tom Regan, "Honey Dribbles Down Your Fur," in *Environmental Ethics: Philosophy and Policy Perspectives*, ed. Philip Hanson (Burnaby, B.C.: Simon Fraser University Publications, 1986), p. 101.

[7] Singer, *Practical Ethics*, p. 92.

limitations of extensionist approaches to the environment. According
to Callicott, first-phase extensionists like Regan and Singer want to
extend moral concern to other sentient animals. Callicott argues that
such first-phase theories are still not environmental ethics, but only
ethics for managing the environment for the use of sentient beings.
The rest of nature would then be "mere means" for the ends of sentient creatures. Callicott argues that these theories are not ethics of the
environment because they do not allow for "direct moral consideration of plants and all the many animals that may not be either sentient, or . . . 'subjects of a life.'"[8]

The disagreements between the extensionists and the holistic theorists are apparent in Callicott's rather harsh critiques of Regan and
Singer and in their replies. Callicott argues that Singer and Regan ignore the "ecological order of nature," which is "premised on one fundamental principle—all life . . . depends on death" (pp. 351–52).
Thus according to the land ethic, "To the extent that the animal liber-
ation/animal rights ethics condemn the taking of life (as a violation of
the rights of a subject of a life or the infliction of pain on a sentient
being), they are irreconcilably at odds with the ecological 'facts of
life'" (p. 352). Thus Callicott claims that Singer and Regan are committed to the absurdity that humans ought to try "to police nature" to
control animal predators who inflict suffering and death on other animals (p. 352).[9]

Callicott is also highly critical of what he calls second-phase extensionists, those who try to extend moral considerability even further
beyond sentient creatures. The second-phase extensionist Kenneth
Goodpaster argues that "the capacity to live—rather than the capacity
to experience pleasure and pain—should be the criterion of moral
considerability."[10] We define a living being in terms of conations, that
is, "inherent tendencies, directions of growth, and natural fulfill-
ments" (p. 353). But here, according to Callicott, the limit of traditional environmental theory is reached, and the theoretical framework
threatens to burst at the seams. The core problem with all these extensionist theories is that they are individualistic, that is, based upon
granting standing to individual members of species, whether human or
nonhuman. But as more and more individuals are granted standing, it
becomes increasingly difficult to reach any practical solution to moral

---

[8] Callicott, "Environmental Ethic," p. 350.
[9] Singer and Regan respond that this is not an implication of their theories. For example, see Singer's *Animal Liberation: A New Ethics for Our Treatment of Animals* (New
York: New York Review, 1975), pp. 238–39.
[10] Callicott, "Environmental Ethic," p. 353.

dilemmas: "Finally, when every living thing is extended equal moral considerability, then moral standing becomes so diluted as to be practically meaningless. Either one must live a life of sacrifice that would make a monk's appear opulent, or one must continually live in a condition of hypocrisy or bad faith" (p. 355). Callicott concludes that Goodpaster's theory "and the set of moral issues that constitute the environmental crisis simply pass one another by, like the two proverbial ships in the night. The lives and deaths of individual shrubs, bugs, and grubs are simply not what the green movement is all about. Rather populations, species, and ecosystems are the objects of environmental concern" (p. 358).

HOLISM: THE LAND ETHIC

Holism centers environmental theory where ecologists claim the center belongs, directly on the environment. As a theory the land ethic has evoked much controversy and more than a little hostility, and can seem quite jarring and unsettling in its challenge to the conceptual frameworks of more traditional theories of value and of the environment. I focus here on the theory as first expounded by Aldo Leopold, and as seen through the eyes of and defended by Callicott.[11] Aldo Leopold in *A Sand County Almanac* propounds a "holistic or ecocentric" environmental ethic.[12] This ethic does not take human or even animal individualistic interests or rights as its focus, but rather centers itself squarely on environmental ecosystems as the prime bearers of inherent value. As Leopold puts it, "A thing is right when it tends to preserve the integrity, stability, and beauty of the biotic community; it is wrong when it tends otherwise" (p. 365). He also claims that "a land ethic changes the role of Homo sapiens from conquerer of the land community to plain member and citizen of it. It implies respect for fellow-members and also respect for the community as such" (p. 364). Callicott justifies his theory on the basis of both ecological and evolutionary Darwinian principles and considerations. While eschewing traditional extensionism, Callicott nonetheless claims, "All con-

[11] Although Callicott at times distances himself from deep ecology, these two approaches sometimes converge, as I show. For expositions of deep ecology, see Arne Naess, *Ecology, Community, and Lifestyle*; "Simple in Means, Rich in Ends," in *Environmental Philosophy: From Animal Rights to Radical Ecology*, ed. Michael E. Zimmerman (Englewood Cliffs, N.J.: Prentice Hall, 1993), pp. 182–92; and "The Deep Ecological Movement: Some Philosophical Aspects," in Zimmerman, *Environmental Philosophy*, pp. 193–212; and Bill Devall and George Sessions, *Deep Ecology: Living as if Nature Mattered* (Salt Lake City: Gibbs Smith, 1985).

[12] Callicott, "Environmental Ethic," p. 361.

temporary forms of life thus are represented to be kin, relatives, members of one extended family. And all are equally members in good standing of one society or community, the biotic community or global ecosystem" (p. 364). This theory holds the prime bearer of value to be the biotic community and other wholes, such as ecosystems and entire species, as opposed to individual members of such species. As Callicott puts it, "The stress upon the value of the biotic community is the distinguishing characteristic of the land ethic and its cardinal strength as an adequate environmental ethic. The land ethic directs us to take the welfare of nature—the diversity, stability, and integrity of the biotic community—to be the standard of the moral quality, the rightness or wrongness, of our actions" (p. 365).

Thus when we are making practical decisions about the environment, we ought to follow the fundamental rule of doing that which will "enhance the diversity, integrity, beauty, and stability of the biotic community" (pp. 365–66). He also argues that nonhuman species as a whole have intrinsic value and that we may be obligated to sacrifice individual members of sentient species in order to save an endangered but nonsentient plant species from extinction.[13]

Callicott, however, tries to embrace both holism and individualism. He appeals to a Humean-Darwinian account of human feelings of benevolence to explain how we can feel sympathy for both individuals and wholes. Darwinian natural selection has operated to select for those members of the human community with more intense and wider-ranging feelings of sympathy. This provides a basis in our feelings for our placing value on the natural environment. He says that both individuals and wholes as such can be "the objects of certain special, naturally selected moral sentiments."[14] Intrinsic value on this model is a bivalent concept grounded on both subjective and objective factors. It "is, as it were, 'projected' onto appropriate objects by virtue of certain naturally selected and inherited intentional feelings, some of which . . . simply have social wholes as their natural objects. . . . Wholes may thus have intrinsic value no less problematically than individuals" (pp. 364–65).

This account of intrinsic value is not without problems of its own, a point to which I shortly turn. But Callicott is at pains to emphasize that in his view the land ethic can accommodate standing for individuals. This is because many of the harshest criticisms of the theory question its commitment to individual members of species and argue

[13] Callicott, *In Defense of the Land Ethic*, pp. 129–55.
[14] Callicott, "Environmental Ethic," p. 364.

that it has inhumane consequences. According to Callicott our moral sentiments can move back and forth between holistic and individualistic objects. He concedes that Leopold's writings stress the holistic aspects and the standing of the biotic community rather than individual members, and he tries to rectify this in his own interpretation of the theory. He wavers on this point, however; and in the end it is not clear where he puts the primary weight.

Callicott is painfully aware of the difficulties engendered by this focus on whole systems. The land ethic has been attacked for its failure to make value distinctions among different species based upon their place on the evolutionary scale. According to this theory Homo sapiens as a species is no more or less valuable than any other species, and its value as a species must be assessed in terms of its impact on the environment. Since this impact has often led to massive environmental destruction, this would seem to place our species low on the value scale and leave little room for pity when war and famine strike. Because there are too many humans causing too much environmental destruction, ecocentrism would seem to require, according to William Aiken, one of its critics, that "massive human diebacks would be good. It is our species' duty to eliminate 90% of our members."[15] According to its critics, ecocentrism also requires a coldhearted attitude toward individual animals of other sentient species if their species overpopulate or are domesticated by humans. This is in conflict with the concerns of those who regard individual animals as having moral standing or rights independent of such ecological considerations and who would not be willing to abandon regard for members of domesticated or overpopulated species. This leads some critics to the conclusion that "the land ethic appears nightmarish in its own peculiar way" (p. 366). It has led Tom Regan to condemn it as "environmental fascism" and L. W. Sumner to call it "dangerous nonsense."[16] In one of his early articles, Callicott seems to play into the hands of critics by aligning himself with and quoting approvingly Garrett Hardin, who claims that "making great and spectacular efforts to save the life of an individual makes sense only when there is a shortage of people. I have not lately heard that there is a shortage of people."[17] Callicott also quotes the deep ecologist Edward Abbey, who states that he "would sooner shoot a man than a snake." Callicott continues, "Abbey may not be simply depraved; this is perhaps only his way of dramatically

[15] William Aiken, quoted by Callicott in "Environmental Ethic," p. 366.

[16] Tom Regan, quoted in Callicott, "Environmental Ethic," p. 366; L. W. Sumner, quoted in Callicott, In Defense of the Land Ethic, pp. 75–99.

[17] Garrett Hardin, quoted by Callicott in In Defense of the Land Ethic, p. 273.

making the point that the human population has become so disproportionate from the biological point of view that if one had to choose
between a specimen of Homo sapiens and a specimen of a rare even if
unattractive species, the choice would be moot."[18]

Callicott tries to counter these condemnations of his theory by ad-
mitting that Leopold and some others have overemphasized its holistic
elements. Callicott claims that the theory is rich and nuanced and can
better accommodate the standing of individuals than these other readings permit. Rather than looking at ethical evolution according to Peter Singer's suggested image of an expanding circle, in which we extend moral standing equally to wider and wider classes of beings, we
should see the evolution of ethics as being like "the image of annular
tree rings in which social structures and their correlative ethics are
nested in a graded, differential system."[19] So it is morally distorted to
suggest a duty to eliminate 90 percent of the current population.

Still, Callicott is far too vague in his guidelines for the obligations of
humans toward the rest of the planet, and he comes up with rules that
could be accepted by most human-centered environmental theorists.
He denies that the land ethic implies that our obligations to humans
are canceled and uses an analogy with human ethics to explain. He
says that simply because "I am now a member of the global human
community and hence have correlative moral obligations to all mankind does not mean that I am no longer a member of my own family
and citizen of my local community and of my country or that I am
relieved of the peculiar and special limitations on freedom of action
attendant upon these relationships" (p. 367). He says that "it is not
unreasonable to suppose that the human community should assume
some obligation and make *some* sacrifice for the beleaguered and
abused biotic community." And further, he claims, "To agree that the
human population should not, in gross and wanton violation of our
humanitarian moral code, be immediately reduced by deliberate acts
of war or by equally barbaric means does not imply that the human
population should not be scaled down, as time goes on, by means and
methods that can be countenanced from a humanitarian point of
view" (p. 368).

Another major criticism of the land ethic is that it lacks a political
and economic theoretical base; it thus incorrectly analyzes the problem of overpopulation as rooted in the moral irresponsibility of Third
World people rather than as having political and economic causes re-

[18] Callicott, *In Defense of the Land Ethic*, p. 27.
[19] Callicott, "Environmental Ethic," p. 367.

lated to poverty and underdevelopment. Ecocentrists proclaim, in writing and conversation, that Third World people are to blame for their predicament because they have too many children and thus it is inevitable that many of them die from disease and malnutrition. They sometimes look on famine, disease, and war as natural phenomena, as nature operating to reduce the human population much as nature operates to reduce the deer population when it gets out of balance with its food supply. But poverty does not reduce overpopulation, it causes it.[20]

This theory faces the major obstacle that many of its actual ethical injunctions remain unpleasant or even horrifying to our usual moral intuitions. Callicott tries to qualify the theory to soften some of its harsher aspects, but he cannot have it both ways. The major strength, as he argues, of ecocentrism is that it places value on wholes, species, and ecosystems equally. When the unpalatable consequences of this are noted, he tries to move back to an approach that accommodates individuals. The result is an uneasy mixture that neither quells the concerns of critics nor provides clear guidelines for cases of conflict. This is a serious obstacle to the plausibility of the theory: if we make ecosystems or the biotic community the central component of the theory, then do we not just have to give up on making individual members of species the focus of moral deliberations? In any case, the land ethic clearly differs from animal rights theories in its acceptance of human killing and eating of other animals and its rejection of the moral duty of vegetarianism for the sake of individual animals, rather than for the sake of purely ecological considerations.

Callicott substitutes the notion of respect for individual members of nonhuman species. Animals may not have rights, but they merit respect, and humans should strive for a "respectful participation in the economy of nature," emulating attitudes of native environmentalism.[21] Humans may consume other animals if they do so with respect. This ethic of respect is undermined, as I will argue, by the attitude toward sport hunting. In trying to put into practice the attitudes of respectful human use of nature, Callicott puts forth two ethical limitations on the use of nature. The first holistic limitation says that human use of the environment should not be destructive but should "enhance the diversity, integrity, stability, and beauty of the biotic community."

---

[20] For a more adequate analysis of this question, see Maria Mies and Vandana Shiva, *Ecofeminism* (Halifax, N.S.: Fernwood Publications; London: Zed Books, 1993), and Tony Killick, *Policy Economics: A Textbook of Applied Economics on Developing Countries* (London: Heinemann, 1981).

[21] Callicott, "Environmental Ethic," p. 369.

The second, individualist limitation is that individual plants and animals used by humans "should be thoughtfully selected, skillfully and humanely dispatched, and carefully used so as to neither waste nor degrade them" (pp. 370, 371). Whether these limitations are sufficient to answer critics remains in question.

ECOCENTRISM AND INHERENT VALUE

Let us turn to some of the deeper theoretical issues concerning value. When we shift our focus from an individualist, consciousness-based framework to one centered on wholes and species and ecosystems, we subtly change the very question we ask about value. Traditional moral theory asks, What sorts of beings have moral standing; that is, what beings ought or deserve to be taken into account for their own sakes when we make moral decisions. In the traditional framework, consciousness is taken as basic because it matters to a conscious being what is done to it. It sounds absurd to say that it matters to a rock or to the Amazon rain forest what is done to it. In the holistic system, however, we ask what sorts of things, conscious or otherwise, have inherent or intrinsic value, since it does not seem absurd to ask whether rain forests have value in themselves. Note that although the question is no longer absurd, it is still a requirement that the case be made that rain forests or other nonconscious things have such value.

How shall we define intrinsic or inherent value? Callicott provides the following: "Something is intrinsically valuable if it is valuable *in* and *for* itself—if its value . . . is independent of any use or function it may have in relation to something or someone else . . . an intrinsically valuable entity is said to be an 'end-in-itself,' not just a 'means' to another's ends."[22]

But the answer of ecocentrism is controversial. There are at least two possible senses of inherent or intrinsic value. In the strong sense, which critic Tom Regan insists upon, if something has inherent value, it must have this value "independently of any valuing consciousness."[23] This reading is reminiscent of the thought experiments used by G. E. Moore in which conscious beings try to imagine extravagantly beautiful worlds entirely without consciousness and ask themselves whether it would be better that such worlds exist even if no conscious being ever had the opportunity to appreciate or value them. This kind of example has led many to think that it is wildly implausible to say that

[22] Callicott, *In Defense of the Land Ethic*, p. 131.
[23] Tom Regan, as paraphrased by Callicott in *In Defense of the Land Ethic*, p. 158.

there can be inherent or intrinsic value independent of an appreciative or valuing consciousness. And yet it is the only way to make the value completely intrinsic to the nonconscious object. Callicott concedes this point and proposes in the place of strong inherent value a second version, which can be called weak inherent value. In this weak sense, an appreciative consciousness is necessary to project value onto a nonconscious object, but the object of appreciation is valued for itself, for properties of its own. As Callicott puts it, "I concede that . . . the *source* of all value is human consciousness, but it by no means follows that the *locus* of all value is consciousness itself or a mode of consciousness like reason, pleasure, or knowledge. In other words, something may be valuable only because someone values it, but it may also be valued for itself, not for the sake of any subjective experience . . . it may afford the valuer" (p. 133).

But Callicott concedes a great deal by this move, without gaining much in return. At the very least we may say that he has admitted the existence of a hierarchy of value, with conscious beings, contrary to the claims of his theory, having greater value than nonconscious valuable things. For although nonhuman species (the immediate subject of his discussion) or ecosystems may be the locus but not the source of value, conscious beings are *both* the source and locus of value, and this confers greater status. Conscious beings, it goes without saying, have the very features that make them valuable in themselves, but they also have the capacities to value and appreciate themselves and others. They project value onto others as well as existing as value unto themselves. The fact that they have this double-impact value is not a small point. For if the light requires to be shined on or projected by others—if value is conferred by others—this places the value of the appreciated object on rather shaky ground. The traditional framework avoids this unhappy consequence by insisting that a being's moral considerability is not conferred by others, although others may acknowledge or refuse to acknowledge their stature. Thus slaves had the same moral considerability as their owners, even though their rights were not recognized and were grossly violated. The great moral revolution and theoretical breakthrough of individual human rights consisted in the insight that all humans were beings who had moral value grounded on features of human nature, and they had this value, based on consciousness, whether or not it was recognized by others. Callicott's projectivist account of intrinsic value moves away from this.

But Callicott's account is suspect beyond this, for if we give up the strong sense of intrinsic value—that is, value independent of consciousness—it is not clear how value resides in the object that is valued. Value is conferred by the lighting up or projection of a human valuer. The features of the object exist whether or not human projectors react with them. The value on this account seems rather more subjectively based in the human valuer than Callicott allows; it is not clear in what sense value resides in the object. At most it resides in a relation between valuing consciousness and features of the object. What resides in the object are natural features that can be picked out by conscious valuers as providing triggers of value. These natural features are good-making properties, but they remain neutral natural properties unless and until value is placed upon them by appreciative valuers. Value remains with consciousness.

Callicott's analysis of value does not establish the conclusion that ecosystems and species are the primary bearers of value. In fact, it shows just the opposite, for since the value of nonhuman species is projected or conferred by individual conscious beings, these latter would seem to be the prime focus of value. Still, his argument does help to establish a weaker conclusion, one that a human-centered moral theorist with serious environmental concerns would welcome. That is, his argument gives grounds for showing why conscious individual humans should respect the natural environment for its own features and not for the sake of any human ends.

This much is welcome. But there are other problems with and constraints on his theory. An environmental ethics needs to give human moral agents some clear guidance in cases of conflict between the very different elements of focus of the theory. Callicott tries to balance concern for wholes, ecosystems, and entire species and their diversity against individual members of human and nonhuman species. Such balancings mirror the complexities and difficulties of the extreme extensionist Callicott ridicules, and yet his theory does little better and has some additional headaches. For example, how exactly do we balance the harm to the environmental system of James Bay, including the possible effects on species, as well as the destruction of the way of life of native people, against the economic benefits of development of hydroelectric power? In this case we may clearly side with the environment and native people, but how have we done the weighing? The case looks more difficult if we try to do a balance when a development project that would clearly benefit a large group of unemployed Atlantic Canadians would endanger a rare plant species. Many such balancings are hard enough from within the traditional framework. But the land ethic calls for a balancing of entirely different orders of things— human and nonhuman individuals versus species and systems—and it calls for the balancing to be done in a new framework, one not dependent on wise or enlightened long-term preservation of nature for human use, but dependent, for example, on a high valuing of endangered species that may be low on the evolutionary scale. When we try to balance such entirely different orders of things, and the values placed on these elements are out of line with our traditional valuings, the weighing problem becomes acute.

A moral theory must do more than give general guidelines. If we need to weigh the value of or interests of such different elements as sentient individuals and nonconscious wholes, we need a theory that has either one fundamental principle to resolve conflicts, or, if there is more than one, we need principles that set out clearly how we prioritize the interests of or value of these different elements consistently.

Recall as well that there are unsettling results of the theory, so the balancings required by this theory may be unclear or they may be horrifying. Callicott's Humean-Darwinian analysis of human sympathy as naturally selected softens the impact. Our sympathy for fellow humans may deflect our judgment that the human population should be reduced drastically. In particular, this sympathy grounds the particular ties we feel with our immediate kin and community.

The Darwinian substructure Callicott appeals to brings out another troubling feature of the land ethic, one that I take up in the context of ecofeminism. The social sentiments that give us special ties to our immediate kin and community coexist with opposite feelings for outsiders—aggression and rage against those from beyond the tribe. Sympathy and aggression are partners in the evolutionary play. And this parallels the ambivalence that ecocentrists and deep ecologists feel toward animals and hunting. Animals are supposed to be killed with respect, but this does not seem to apply to sport hunting, which is celebrated by these theorists. This is one of the firmest dividing lines with ecofeminism.

Thus there are serious problems with the land ethic: an analysis of value that does not support the value claims of the theory; a lack of guidance on how weighings of very different elements are to be carried out; and decisions that are troubling at best, horrifying at worst. I turn now to one of the newest alternatives, ecofeminism.

ECOFEMINISM

Ecofeminism uses the resources of recently developed feminist theory to shed light on and build theories of the environment. It is often noted that ecocentrism, deep ecology, and ecofeminism share some basic tenets and part company in basic ways.[24] All are ecological approaches and as such try to get beyond purely human-centered attitudes toward the environment. All hold in their gaze the flux and flow of holistic systems as being centers of value. But as Regan and others have pointed out, a major difference is that ecofeminism, as a feminist theory of the environment, points to the major cause of the ills of our planet and our human culture as being, not merely "human-centeredness (not anthropocentrism) but male-centeredness (androcentrism)" (pp. 135–36). Or put another way, the source of the problem is patriarchy, not humanism.

Ecological feminism begins from the assumption that "the domination of woman was the original domination in human society, from which all other hierarchies—of rank, class, and political power—flow."[25] Karen Warren claims that "ecological feminism is the position that there are important connections—historical, experiential, symbolic, theoretical—between the domination of women and the domination of nature, an understanding of which is crucial to both feminism and environmental ethics."[26] The unity of women and nature is taken to be the basis of patriarchal domination of both. Warren says that the problem is oppressive conceptual frameworks that function to maintain domination and subordination. She describes these as follows:

> There are three significant features of oppressive conceptual frameworks: (1) value-hierarchical thinking, i.e., "up-down" thinking which places higher value, status, or prestige on what is "up" rather than on what is "down"; (2) value dualisms, i.e., disjunctive pairs in which the disjuncts are seen as oppositional (rather than as complementary) and exclusive (rather than as inclusive), and which place higher value (status, prestige) on one disjunct rather than the other (e.g., dualisms which give higher value or status to that which has historically been identified as "mind," "reason," and "male" than to that which has historically been identified as "body," "emotion," and "female"); and (3) logic of domination, i.e., a structure of argumentation which leads to a justification of subordination. (p. 128).

Ecofeminism sees all life as "an interconnected web, not a hierarchy."[27] Feminist theory often challenges dualisms or splits, and ecofeminism

---

[24] See Tom Regan, "Irreconcilable Differences," in *The Thee Generation* (Philadelphia: Temple University Press, 1991), pp. 125–42.

[25] Ynestra King, "The Ecology of Feminism and the Feminism of Ecology," in *Healing the Wounds*, ed. Judith Plant (Toronto: Between the Lines, 1989), p. 24.

[26] Karen J. Warren, "The Power and the Promise of Ecological Feminism," *Environmental Ethics* 12 (Summer 1990): 126.

[27] King, "Ecology of Feminism," p. 19.

challenges the nature-culture dualism. Patriarchy is based on this form of dualism, and in building up Western culture and industrial societies, "nature became something to be dominated, overcome, made to serve the needs of men" (p. 20). Men could have unlimited use of and control over nature, and any problems that arose could be solved by male technological prowess. Thus nature, like women, in the patriarchal attitude became "other, something . . . to be objectified and subordinated" (p. 21). Simone de Beauvoir eloquently expresses the ambivalent feelings man has for nature: "He exploits her, but she crushes him, he is born of her, and dies in her; she is the source of his being and the realm that he subjugates to his will; nature is a vein of gross material in which the soul is imprisoned, and she is the supreme reality. . . . Now ally, now enemy, she appears as the dark chaos from whence life wells up."[28] King suggests that we choose consciously "not to sever the woman-nature connection" but to value it (p. 23). This is one of the entry points that this theory has for a valuing of nature in itself. As King puts it: "Ecological science tells us that there is no hierarchy in nature itself, but rather a hierarchy in human society that is projected onto nature. . . . We live on the earth with millions of species, only one of which is the human species. Yet the human species in its patriarchal form is the only species which holds a conscious belief that it is entitled to dominion over all the other species and over the planet" (p. 24).

Ecofeminism also shifts the focus of the discussion. Instead of asking why we ought to value something, or some things more than other things, ecofeminism tries to undercut the question by claiming that hierarchies, including value hierarchies, and hierarchical conceptual frameworks are part of patriarchal thinking and we should value all aspects of life and the environment. We should conceive of ourselves and other elements of the natural world not as individuals, but as a web of elements connected by relationships. The key question, built on feminist ethics, becomes not just what we can value but what we can care for and care about. We need to reconceptualize ourselves and how we relate to the rest of the environment. All parts of an ecosystem are valuable, and we must put the emphasis on "the independent value of the integrity, diversity, and stability of ecosystems, or on the ecological themes of interconnectedness, unity in diversity, and equal value to all parts of the human-nature system."[29]

---

[28] Simone de Beauvoir, *The Second Sex*, p. 144, quoted in King, "Ecology of Feminism," p. 21.

[29] Karen J. Warren, "Feminism and Ecology: Making Connections," *Environmental Ethics* 9 (Spring 1987): 7, 10.

For a Darwinian account of human feeling, ecofeminism substitutes a feminist ethic of care. This underlying view of the place and content of human feeling is one of the clearest examples of the irreconcilable differences between ecofeminism and ecocentrism and deep ecology. It replaces a masculinist moral psychology with a feminist one in which feelings of love and care are central and there is no place for aggression. As part of a needed transformation we must bring into a new balance certain values of care, friendship, reciprocity in relationships, and appropriate trust, all of which have been undervalued in traditional frameworks.

Other traditional frameworks have asked why we ought to value the natural environment. Ecofeminism starts from the position that all of nature has value and we should not impose value hierarchies upon it. Rather we need to acknowledge that our traditional frameworks have encouraged us to think of nature as there for our ends, to be used and exploited. If we work to change the ways of thinking that have encouraged the exploitation of nature, we will naturally and spontaneously substitute an attitude of respect for nature. By stressing an ethics grounded on caring, nurturing, relationships, and connectedness, ecofeminism tries to bring into proper balance and perspective the traditional moral framework, which relies on arguments and reason to establish the value of animals or ecosystems.

Another challenge is to what is alleged to be the patriarchal sense of the self and the kinds of ethical arguments and theories that accompany such a sense of self. Ecofeminists want to move away from a conception of persons as atomized individuals whose rights or interests are in conflict or competition with one another, and to move toward a conception of persons as part of an interconnected web tied together by relations of love and appropriate trust. This puts great weight upon the socialization of members of the moral community to be trustworthy moral agents. This idealized moral community is one "in which individuals are what they are in virtue of the trust, love, care, and friendships that bind the community together."[30]

The feminist care ethic that underlies ecofeminism is not without its difficulties. Several commentators have raised concerns that this care ethic actually can be regressive and has the potential to reinforce harmful stereotypes of women and impede their ability to overcome oppression. They warn that it is important to examine and scrutinize our caring relations. Christine Cuomo, for example, says that "femi-

---

[30] Jim Cheney, "Eco-Feminism and Deep Ecology," *Environmental Ethics* 9 (Summer 1987): 129.

nists must ask if caring for other particular beings or objects is a good activity to engage in when one is trying to free oneself from a subordinate social position . . . female caring and compassion for oppressors are cornerstones of patriarchal systems. Women have forgiven oppressors, stayed with abusive husbands and partners, and sacrificed their own desires because of their great ability to care for others."[31] These and similar concerns are echoed by other authors.[32]

Ecofeminism also takes a bold and radical approach to the problems I have raised concerning conflict resolution or how we weigh the value or interests of individuals against the wholes of species or systems. Karen Warren claims that it is a positive virtue of an ethical framework if "conflict is unresolvable, at least by ordinary hierarchical, adversarial means."[33] This gives us the impetus to take seriously "the call by feminists for alternative methods of conflict resolution" such as "consensual decision making" (p. 132). Of course, this claim relies on the background of the expanding literature on feminist ethics of care.

I concur that feminist strategies to confront attitudes that view nature as there for human exploitation, as well as strategies to increase human respect for the rest of the planet, are a welcome addition to the environmentalist program, and consensual decision making to resolve ethical dilemmas is also welcome when it works. But when consensus does not work, and we are faced with genuine and painful conflict, then ecofeminism provides little guidance in particular cases if the conflicting claims are all seen as being of equal value.[34] Other feminists have offered suggestions how we may make practical moral decisions when we value equally and base our approach on an ethics of care. As Jim Cheney puts it, "It is the contextual nature of ethical decisions, the fact that they occur in the context of web-like relations of care and responsibility—together with the fact that we cannot care effectively in *all* cases—that makes for differential moral regard. It is not . . . a difference in value, worth, or rights . . . that calls for differential moral

---

[31] Christine Cuomo, "Unravelling the Problems in Ecofeminism," *Environmental Ethics* 14 (Winter 1992): 354–55.

[32] See also Janet Biehl, *Finding Our Way: Rethinking Ecofeminist Politics* (Montreal: Black Rose Books, 1991), pp. 140–47; Jean Grimshaw, *Philosophy and Feminist Thinking* (Minneapolis: University of Minnesota Press, 1986), pp. 182–83; Sarah Lucia Hoagland, "Some Thoughts about 'Caring,'" in *Feminist Ethics*, ed. Claudia Card (Lawrence: University Press of Kansas, 1991), pp. 246–63; and Wendy Donner, "Self and Community in Environmental Ethics," in *Ecofeminism*, ed. Karen J. Warren (Bloomington: Indiana University Press, forthcoming).

[33] Warren, quoted by Cheney, in "Eco-Feminism," pp. 131–32.

[34] Janet Biehl has some harsh criticisms of ecofeminist consensus decision-making processes, which she argues are undemocratic. See Biehl, *Finding Our Way*, pp. 136–42.

regard. The limits of ethical considerability are the limits of one's . . .
ability to care and show concern."[35]

Yet this move should put us on the alert. Have we not here ex-
changed one set of difficulties and perplexities for another set just as
or unsettling if not more? Can one who has serious concerns for the
well-being of the environment trust the human capacity to care
more than the human capacity to reason and value? It is well to ask
for greater empathy with the environment. But Tom Regan has a
powerful point to make here. He says, "ecofeminists must address
the problem of how an ethic of care can overcome anthropocentric
limits—that is, how nurturing care gets extended to wolves and coy-
otes, deer and whales, for example, not just family and friends."[36]
Feeling alone will not suffice, for feelings can be quite arbitrarily
applied. Janet Biehl points out that "individuals may easily start or
stop caring. They may care at their whim. They may not care
enough. They may care about some but not others. Lacking an institutional form and dependent on individual whim, 'caring' is a slen-
der thread on which to base an emancipatory political life."[37] What
is needed is the balance of reason in the form of an argument about
consistency. We care about the pain of those close to us because of
what pain and suffering are. Consistency leads us to care about suffering of any sentient creature. Regan says, "Whether I care or not, I
ought to. And how I know that I ought to care is by recognizing that
logic leads me to this conclusion."[38]

What is striking about caring and empathy in the human population
is just how limited in scope and arbitrarily applied it can be. The heart
on its own cannot supply the need. Heart and head are both needed.

The ethic of caring is, however, one way in which ecofeminism is
superior to ecocentrism and deep ecology. Callicott calls on resources
of human feeling as a central feature of his theory. But Callicott's
Darwinian and sociobiological sentiments are markedly different
from, and I contend inferior to, feminist care and empathy for environmental concerns. The sociobiological account underlying Callicott's argument is masculinist, not feminist. It gives equal place to
human feelings of aggression and bloodlust and to sympathy. This
becomes chillingly clear when we examine the deep ecological and
ecocentrist attitudes to sport hunting. They extol the virtues of hunting even in cases in which it is not called for to meet sufficiency needs

[35] Cheney, "Eco-Feminism," p. 138.
[36] Regan, "Irreconcilable Differences," p. 139.
[37] Biehl, *Finding Our Way*, p. 150.
[38] Regan, "Irreconcilable Differences," p. 140.

or to control the animal population. In their eyes it is truly a blood sport. Randall Eaton observes that "the hunter's feeling for his prey is one of deep passion and ecstacy. . . . The hunter loves the animal he kills."[39] Aldo Leopold says, "The instinct that finds delight in the sight and pursuit of game is bred into every fibre of the human race" (p. 137).

The feminist ethic of care and its feelings of sympathy and love seem to be a much more authentic base for a genuine respect for the environment and nonhuman animals. Death may be a necessary part of nature, as it is of human life, but we are not called upon to revel in it or feel ecstatic about it.

Ecofeminism seems to have many problems in providing guidance for practical actions with regard to the environment. Still, this theory was developed very recently, and thus charity should be extended to allow for time to work out its program. It seems to be a more promising avenue for working out a basis for motivating humans to respect the environment, since it has no place for sociobiological feelings of aggression. Respect and aggression do not mix. Those who love what they kill or love to dominate the objects of their love are alienated from the object of their love. Ecofeminism therefore does not have room for the troubling or horrifying consequences of ecocentrism and deep ecology. Ecofeminists are also concerned about issues of poverty, race, and class and have at their disposal a much sounder socioeconomic analysis of the causes of world poverty. They do not welcome war and famine as natural means of culling the human race, but view the death of their sisters and brothers in impoverished nations as a great tragedy that could be controlled and ended by appropriate social and political measures.[40]

DEEP ECOLOGY AND FEMINISM

One issue that has given rise to considerable controversy in environmental ethics is the debate between proponents of ecofeminism and deep ecology over the appropriate account of the self in environmental philosophy, as well as the appropriate account of self and other, in particular self and nature.

The most prominent deep ecology theorist is Norwegian philosopher Arne Naess. He contrasts what he calls shallow ecology, which

[39] Quoted by Regan in "Irreconcilable Differences," p. 136.
[40] For an ecofeminist theory that has a compelling analysis of these issues, see Mies and Shiva, *Ecofeminism*.

concentrates on human-centered reformist approaches, with deep ecology, which is "proposing a major realignment of our philosophical worldview, culture, and lifestyles, consistent with the new ecological perspectives."[41] Naess and George Sessions have advanced an eight-point platform that they believe supporters of deep ecology as a movement embrace.[42] The key philosophical principles underlying this platform are summed up by Sessions as "an awareness of the internal interrelatedness of ecosystems and the individuals comprising them; ecological egalitarianism (ecocentrism as opposed to anthropocentrism); an appreciation of diversity, symbiosis, and ecological complexity.[43] These may be further compressed, for our purposes, into the following principles: ecocentrism, the rejection of anthropocentrism, and self-realization. Here I propose to set aside further scrutiny of the first two and to focus on the analysis of the self.

Naess explains the norm of self-realization:

> I do not use this expression in any narrow, individualistic sense. I want to give it an expanded meaning based on the distinction between a large comprehensive Self and a narrow egoistic self as conceived of in certain Eastern traditions of *atman*. This large comprehensive Self . . . embraces all the life forms on the planet . . . together with their individual selves.
> . . . This leads to a hypothesis concerning an inescapable increase of identification with other beings when one's own self-realization increases. As a result, we increasingly see ourselves in other beings, and others see themselves in us. In this way, the self is extended and deepened as a natural process of the realization of its potentialities in others.[44]

Deep ecologist Warwick Fox puts forward this version of the central claim of deep ecology: "We can make no firm ontological divide in the field of existence . . . there is no bifurcation in reality between the human and nonhuman realms . . . to the extent that we perceive boundaries, we fall short of deep ecological consciousness."[45]

Deep ecology and ecofeminism both take seriously the importance of a new account of the self in harmony with environmental ethical theories. They both also share a common analysis of the problems with the self posited in more traditional, anthropocentric theories—

[41] George Sessions, "Introduction," in Zimmerman, *Environmental Philosophy*, p. 162.
[42] Naess, "The Deep Ecological Movement," p. 197.
[43] Sessions, "Introduction," p. 163.
[44] Naess, "The Deep Ecological Movement," p. 209.
[45] Warwick Fox, as quoted by Val Plumwood, in "Nature, Self, and Gender: Feminism, Environmental Philosophy, and the Critique of Rationalism," in Zimmerman, *Environmental Philosophy*, p. 293.

that is, that this self is an egoistic self, split off from, separated and alienated from nature. They both attempt to overcome these deficiencies and alienations by outlining what they view as more appropriate accounts of the self. But at this point they part company. Val Plumwood's account of the ecofeminist self-in-relationship is arrived at partly through her critique and rejection of several alternative accounts of the deep ecological self. She labels the problem of separation from nature "the discontinuity problem" (p. 293), but she claims that deep ecology's attempts to overcome it do not succeed: "Deep ecology locates the key problem area in human-nature relations in the separation of humans and nature, and it provides a solution for this in terms of the 'identification' of self with nature" (p. 293). But identification is a vague concept, and so Plumwood examines three different versions, which have been held by ecologists: the indistinguishability of self, the expansion of self, and the transcendence of self. Her critiques of all three are illuminating, but here I have space to examine only the indistinguishability account of identification.

She sets out the indistinguishability account of identification of self with nature: "The indistinguishability account rejects boundaries between self and nature." To resolve the problem of discontinuity between humans and nature, this deep ecology view "proposes to heal this division by a 'unifying process,' a metaphysics that insists that everything is really part of and indistinguishable from everything else." According to Plumwood, this move to overcome the problem by denying all distinctions is too drastic. For one thing, it allows no room to scrutinize the *kind* of relation humans have to nature, for "the situation of exploitation of nature exemplifies such unity equally as well as a conserver situation" (p. 293–94). This view of the self also has the sort of clear dangers that I have already pointed out in my discussion of care ethics—for boundaries, not self-merging, are needed in healthy caring relations. Plumwood stresses that "we need to recognize not only our human continuity with the natural world but also its distinctness and independence from us and the distinctness of the needs of things in nature from ours. The indistinguishability account does not allow for this, although it is a very important part of respect for nature and of conservation strategy" (p. 295).

Plumwood tries to counter the dangers she sees in the identification approach of deep ecology by arguing for a relational account of the self that does not deny the independence of the other, including nature. Her solution, the self-in-relationship, "enables a recognition of interdependence and relationship without falling into the problems of indistinguishability, that acknowledges both continuity and difference" (pp. 301–2). Karen Warren provides a succinct comment about the difference between the self of deep ecology and the self of ecofeminism when she claims that an ecofeminist perspective "presupposes and maintains difference—a distinction between the self and other." Further, in her account, "one knows 'the boundary of self,' where the self—the 'I' . . . leaves off . . . There is no fusion of two into one, but a complement of two entities acknowledged as separate, different, independent, yet in relationship."[46]

This debate is ongoing and unresolved; for the moment I can only record my agreement with the need for distinct boundaries as well as relations, as ecofeminists maintain, rather than the fusion and merging of the deep ecological self.

The pieces of the puzzle are there. Fundamentally, I believe, we must have an environmental ethic that accords an appropriate and central place to sentient individual members of human and nonhuman species. I have also searched for an analysis that extends value beyond consciousness to include individual members of nonconscious species, entire species, ecosystems, and the whole biotic community. The arguments for inherent value beyond consciousness I have found less compelling, especially in cases where serious interests of conscious beings may be endangered. Nevertheless, I have come up with arguments and perspectives that give insight into why we ought to grant respect to these other elements of nature, including respect for features of their own, apart from their impact on human interests. Part of the way out of the environmental crisis may well involve, as both ecofeminists and deep ecologists claim, a new account of the self at the core of an environmental theory stressing harmony with and interconnectedness with nature. Perhaps their differences over the details of this account are overshadowed by their common insights into the problem. While we want a serious and theoretically sound environmental ethics, one that gives practical guidance on painful matters, what we also need are ways to motivate humans to respect the natural on its own terms. Thus the appeal must be to both our hearts and our heads.

[46] Warren, "The Power and the Promise," p. 137.

4

Societies in Space and Place

IAIN WALLACE and DAVID B. KNIGHT

DIFFERENTIATION both in the world's natural environment and in the cultural contexts of human interaction with it challenges the concepts of justice invoked in contemporary efforts to come to terms with environmental change at various geographical scales. Between universalistic "views from nowhere" and ideological parochialisms there exist modes of thinking and being that are rooted in space and place yet capable of contributing to a global ethic of responsible action governing human transformations of the natural world.

TAKING "EARTHINESS" SERIOUSLY

The Earth is not uniform. Nearly three-quarters of its surface is covered by the oceans. The rest is divided into distinct continental blocks, irregular in shape, and located at different latitudes. The elevation of the land above sea level varies greatly and so does the distribution of major geomorphic features such as river systems. There are significant differences across the globe in climate and, largely as a direct result, in the prevailing types of life forms. Even before the presence and impacts of human societies are considered, the vast range of dynamic ecosystems makes the Earth's surface a marvelously varied phenomenon. This natural "given," this differentiation, lies beyond the domain of concepts of justice, but the biodiversity to which it gives rise is increasingly acknowledged as intrinsically valuable.

"Traditional" or preindustrial human societies are intimately shaped by the biophysical potentials of their encompassing ecosystems, yet they also clearly display varieties of technological adaptation and cultural constructions of meaning. The human capacity to shape, or at least make a viable adaptation to, a variety of natural environments takes us into the world of culture without forsaking nature completely.[1] The differentiated character of the Earth's surface (more narrowly its land surface, including here the immediate subsurface and atmospheric features) determines the geographical variations in its capacity to support life. One measure of this capacity is the net primary organic productivity (NPP) of each regional ecosystem—essentially its capacity to convert incoming solar energy (directly or indirectly received) into terrestrial vegetation, on which "higher" life forms ultimately depend.[2] By this measure, a tropical rain forest is intrinsically more productive (has a higher mean NPP) than a boreal forest, a high-altitude grassland, or a hot or cold (tundra) desert, for instance. Other things being equal, a "traditional" or preindustrial society situated in an ecosystem of higher NPP will enjoy a higher sustained material standard of living than one situated in an ecosystem of lower NPP, as a direct result of the underlying nonisotropic natural conditions. Economic *inequality* between societies in these circumstances is not, prima facie, *inequity*. The Inuit of arctic Canada have a more constrained and "harder" life than the Polynesians of Tonga, but like indigenous peoples the world over, both societies have developed a concept of the "good life" and are "at home" in their geographical settings.[3] Although afforded even less protection than biodiversity in contemporary regimes of global and national governance, the diversity of indigenous cultures is increasingly being acknowledged as an intrinsic good and its preservation a matter of justice. Moreover, traditional ecological knowledge has a newly respected instrumental value.[4]

The propensity of certain ecosystems ("bioregions") to sustain a rel-

[1] Environmental determinism, a fascinating but now discredited reading of human accommodation to the variety of natural biophysical conditions, was well expressed by Ellen Churchill Semple, *Influence of Geographic Environment* (New York: Holt, 1911). It was strongly countered by the cultural geographical perspective in, for example, George Carter, *Man and the Land* (New York: Holt, Rinehart & Winston, 1964). I. G. Simmons, *Changing the Face of the Earth: Culture, Environment, History* (Oxford: Basil Blackwell, 1989), provides a recent integrative account.

[2] S. R. Eyre, *The Real Wealth of Nations* (London: Edward Arnold, 1978), provides global estimates of NPP by major bioregion. From these, he derives estimates of the potential productivity of human foodstuffs, expressed in per capita annual nutrition requirement units.

[3] See Yi-Fu Tuan, *The Good Life* (Madison: University of Wisconsin Press, 1986).

[4] See Julian T. Inglis, *Traditional Ecological Knowledge: Concepts and Cases* (Ottawa: International Development Research Centre, 1993).

atively high NPP provides their human inhabitants with the basis on
which to develop a more complex material culture—to more readily
accumulate a surplus above subsistence needs. We would argue that it
is at this point, when a society must decide how to consume or deploy
that surplus, that we enter decisively the domain of moral judgment.
Thus any attempt to resolve questions of social justice surrounding
global (or more localized) environmental change needs to take seriously the underlying geographical variety of the Earth's potential to
support a human society. Within the observable disparities of living
standards enjoyed by various societies around the globe, there are
both natural and culturally induced sources of difference. It is important that we acknowledge the contribution of each and consider the
bearing each has on our concepts of justice. What should we value in
the characteristics of and the extent of diversity among human populations?

"Modern" history sharpens this question. An industrial revolution
has freed most societies from exclusive dependence on renewable environmental resources. This, together with the creation of an international state system, the rise of an increasingly integrated global capitalist economy, and the growth of a hegemonic culture proselytizing the
values of modernity and rising material consumption—an ensemble of
forces and institutions grounded in, and hence shaped by, the experience of Western Europe and its transatlantic offspring—has irrevoca-
bly broadened the scope of our concern. If the varied environmental
productivity (organic NPP, to which can be added measures of the
abundance of nonrenewable "resources") accessible to societies in different parts of the world is, in itself, unproblematically "natural," the
same cannot be said of the scale of the contemporary human appropriation of it, nor of the ways in which particular "modern" societies
and institutions have gained control over specific portions of it.[5] Dominant societies the world over have guilty histories, which have created

[5] Eyre, *Real Wealth of Nations*, expresses the current monetary value of each state's
mineral resource production in terms of per capita annual nutrition requirement units.
He then ranks states in terms of these "nutrition units," based on their combined organic and mineral-equivalent primary productivity, and thus manages to compile an
approximation of human "carrying capacity" for each state. For a discussion of the
World Bank's more comprehensive experiment in reckoning of national wealth, see John
O'Connor, *Monitoring Environmental Progress: A Report on Work in Progress* (Washington, D.C.: World Bank, 1995). The history of European states' incorporation of the
territory and natural resources of most of the rest of the world is detailed in Immanuel
Wallerstein, *The Modern World System* (New York: Academic Press, 1976, 1980,
1989); the history of their ecological imperialism, in A. W. Crosby, *Ecological Imperialism: The Biological Expansion of Europe, 900–1900* (New York: Cambridge University
Press, 1986).

distinctive geographies, not least of environmental change. The consequences reflect the spatialization of power, globally, nationally, and locally. Responding to the challenges of these impacts with an eye to claims of justice demands an equal sensitivity to the contexts of space and place.

GEOGRAPHY, SOCIAL THEORY, AND JUSTICE

The ethical and social theories that have dominated modern Western culture have been fundamentally individualistic and universalistic.[6] The rational self, divorced from his social and environmental context (the discounting of relatedness signifies an unambiguously masculine subject), makes choices whose formal purity is uncontaminated by particularistic associations to place (whether natural setting or human community) or by the constraints of space (which includes distance and differential accessibility). It is worth exploring more fully the long-standing neglect of these concepts in discourse about society, for it will help identify how they enrich our grasp of what social justice entails.[7]

One consequence of the search by natural scientists for what were believed to be universally invariant characteristics of phenomena was to abstract from spatial coordinates: *where* observations were made did not matter, because geographical location per se made no difference to the results. For disciplines that modeled their research paradigms on physics, neither locations (the elements of "space") nor the complex particularities of "place" aroused much intellectual interest. The development of social science within the same epistemological framework encouraged continued neglect of these dimensions of reality. With a few notable exceptions, economists have ignored location and the environmental parameters of places as analytically useful variables, and sociological theory has tended to associate the significance of space and place with conventionally defined entities, such as the city, rather than seeing them as categories constitutive of all social

---

[6] Universalizing is not a feature of "advanced" Western societies alone. Most societies, including preliterate hunter-gatherers, see themselves at the "center" of the "world" (even of the cosmos) and hence refer to or accept themselves simply as "*the* people." It is instructive that, among others, ancient Greeks, Chinese, and Christians all identified specific places as the "navel of the earth." See Yi-Fu Tuan, *Topophilia* (Englewood Cliffs, N.J.: Prentice-Hall, 1974), especially pp. 30–44.

[7] J. A. Agnew, "The Devaluation of Place in Social Science," in *The Power of Place: Integrating Sociological and Geographical Imaginations*, ed. J. A. Agnew and J. S. Duncan (Boston: Unwin Hyman, 1989), pp. 9–29.

phenomena.[8] It has been left to the "lesser" disciplines of geography and anthropology to keep alive interest not just in putative universals of human experience but also in the particularities of environmental and social context. The past decade has seen, however, a broader acknowledgment of the significance of space and place within social theory.[9] With it has come much wider acceptance that "geography matters," not least in the achievement of social justice:[10] that the Earth is not simply a passive stage on which human dramas (or invariant processes) are played out, but that place—the particular social and natural environment in which actors are set—is a formative influence on who they are, individually and as a community, on how they live, and on how they think.

Two relevant examples of the geographical situatedness of ostensibly generalized (universalistic) theorizing are John Rawls's *Theory of Justice* and Herman Daly and John Cobb's *For the Common Good*.[11] Rawls focuses on the workings of democratic polities, which comprise a substantial fraction of the world political map, but his work promotes an image of "the good life" that is, as Gordon Clark has argued, "quite American."[12] The rights-holding individual is the central moral agent, whose freedom of action is to be circumscribed neither by bonds of obligation to an ethical community nor by the potentially coercive demands of the state. This theory of justice is much more in tune with the cultural context of the United States than, say, that of western European social democracy; and indeed, Rawls has subsequently acknowledged that his work should be seen as politically (and, we would argue, geopolitically) embedded rather than essentially

[8] The integration of economic and spatial concepts achieved by authors such as August Losch and Walter Isard never entered mainstream economic theory. An economics that takes the biophysical constitution of the Earth seriously is only recently emerging, building on pioneer work by such as Nicholas Georgescu-Roegen and Herman Daly. The human ecology school of urban sociology, which developed at Chicago in the 1920s, was certainly grounded in the particularities of that city's evolution, but its practitioners tended thereafter to overgeneralize from the Chicago model.

[9] See, for example, David Harvey, *The Condition of Postmodernity* (Oxford: Basil Blackwell, 1989); Edward W. Soja, *Postmodern Geographies: The Reassertion of Space in Critical Social Theory* (London: Verso, 1989); and Anthony Giddens, *The Consequences of Modernity* (Stanford: Stanford University Press, 1990).

[10] See Doreen Massey and John Allen, eds., *Geography Matters!* (Cambridge: Cambridge University Press, 1984), and David M. Smith, *Geography and Social Justice: Social Justice in a Changing World* (Oxford: Blackwell, 1994).

[11] John Rawls, *A Theory of Justice* (Cambridge: Harvard University Press, 1971); Herman E. Daly and John B. Cobb Jr., *For the Common Good: Redirecting the Economy toward Community, the Environment, and a Sustainable Future* (Boston: Beacon Press, 1989).

[12] Gordon L. Clark, "Making Moral Landscapes: John Rawls' Original Position," *Political Geography Quarterly* 5 (1986) Supplement: S149.

metaphysical (and hence grounded nowhere).[13] Moreover, in abstracting from issues of *inter*national justice, Rawls implicitly assumes the normality of a relatively self-sufficient national economy—which is what the United States has historically been, unlike most other states. His projection of the "original position" model to the international scale not only nullifies real historical contingencies, but also those of geographical position and environmental endowment. To argue that "the basic principle of the law of nations is a principle of equality" comes close to invoking a biophysically isotropic globe as a background condition—unless he is prepared to grapple concretely with the environmental implications of "justice" derived from putting the United States and Burkina Faso, for instance, behind the same veil of ignorance.[14]

Although Daly and Cobb's formulation of "the common good" makes less of a pretense to universality than Rawls's "justice," they nevertheless demonstrate an insensitivity to the particularity of the "place" out of which they write. In arguing that the well-being of global society lies in effecting the "closure" of national economies in order to foster self-sufficiency in "communities" defined at a variety of subnational scales, they too are reflecting the historical experience of the United States, which can more readily pursue protectionist, quasiautarkic policies than can any other member state of the OECD. Even where a national "community" exists, states find increasing difficulties in nurturing it in the face of disruptive and largely uncontrollable external pressures, notably those emanating from the global economy. A country's geographical scale, economic structure, and degree of environmental resource–based self-sufficiency all have a major influence on the capacity of its government to insulate citizens from forces (such as industrial restructuring and the declining terms of trade of primary commodities) that by impacting places and classes differentially tend to polarize society. The ability to modulate the advantages of economic openness and closure that accrues to the United States, as much from its varied domestic geography and resource endowment as from its geopolitical leverage, is a rare asset. The potential benefits, as well as undoubtedly the costs, of economic specialization within a system of international "free" trade, are more readily apparent in smaller countries.

Critiquing the vulnerable pretensions of universalistic discourse and

[13] See John Rawls, "Justice as Fairness: Political Not Metaphysical," *Philosophy and Public Affairs* 14 (1985): 253–51.

[14] See Rawls, *Theory of Justice*, p. 378, and also R. D. Winfield, *Freedom and Modernity* (Albany: State University of New York Press, 1991), chap. 16.

advancing the claims of more contextualized (localized) knowledges, is not, of course, an unalloyed "postmodern" advance. It raises the specter of unprincipled relativism and potentially antagonistic parochial ideologies, neither of which are likely to promote just and judicious responses to the threats of global environmental change.[15] But normative philosophies based on atomistic individualism or on an unquestioned acceptance of the state as the prime political agent are not likely to either. It is at the intermediate scale of the "community" or substate society that much contemporary interest is focused, and at which many of the challenges framing ethical human responses to society's environmental impacts can be most clearly articulated.

ETHICAL DIMENSIONS OF GEOGRAPHICAL SITUATEDNESS

As critics of the atomistic individualism of classical political and economic theory have long argued, the necessary situatedness of every person in a biophysical and social system, far from being a trivial incidental, is fundamentally constitutive of who they are. Our understanding of the world is shaped by the environments, natural and cultural, in which we live. In premodern societies, for most (but far from all) people, the association between place, livelihood, and life-path was stable and enduring, fostering the emergence of local communities of distinctive social structure and mores. But we would argue that even today, in societies marked by infinitely greater residential and occupational mobility and personal networks of association that are potentially global in scope, very few people are radically "placeless."[16] They reside in particular localities, which may or may not resemble "communities" in the traditional sense but are imbued with particular characteristics of the natural and built environment and an identifiable moral quality, which reciprocally influence the worldview and values of their inhabitants. People are shaped by "where they come from" in more than just a colloquial sense. Hence Michael Sandel, articulating the communitarian critique of a rights-based liberalism, can claim that "if we are partly defined by the communities we inhabit, then we must also be implicated in the purposes and ends characteristic of those communities . . . the story of my life is always embedded in the story of those communities from which I derive my identity—whether fam-

[15] See Philip Howell, "The Aspiration towards Universality in Political Theory and Political Geography," *Geoforum* 25 (1994): 413–27.

[16] See J. Nicholas Entrikin, *The Betweenness of Place: Towards a Geography of Modernity* (Baltimore: Johns Hopkins University Press, 1991).

ily or city, tribe or nation, party or cause. On the communitarian view, these stories make a moral difference, not only a psychological one. They situate us in the world, and give our lives their moral particularity."[17]

Of course, the "moral particularity" of a community that is geographically bounded is equally open to interpretation as a parochial or regional ideology. As Alexander Murphy argues: "All regional formations carry with them particular conceptions that influence activities and practices. As social constructions, regions are necessarily ideological and no explanation of their individuality or character can be complete without explicit consideration of the types of ideas that are developed and sustained in connection with the regionalization process."[18] Insofar as a community's identity is defined through a process of denigrating a contrasted and distant "other," its reactionary character is readily apparent. Even the renewed concern for "place" that is part of postmodern sensibilities has been judged to be no more than an ideological mask for socially regressive nostalgia and aestheticization.[19] But the specificity of the local need not be reactionary: indeed, if social justice is to be made concrete in the lives of men and women, it has to engage the particularity of the places in which those lives are embedded.[20] Ethical solutions to the inevitable conflicts that arise from a clash of the particular interests of territorially based communities cannot be derived exclusively from a universal proposition, a utopian "general" interest, or a counterfactual "initial position" that stands "above" (ungrounded in) the partisan positions and the valuing that shapes them. As the Arab-Israeli peace process illustrates, one is required to grapple concretely with those constitutive differences of place and experience (involving the guilty history of one or more protagonists) and to effect a reconciliation of claims to fairness that can be embedded tangibly in the biophysical environment and social geography of particular territories.[21]

[17] Michael Sandel, "Introduction," in Liberalism and Its Critics, ed. Sandel (Oxford: Basil Blackwell, 1984), pp. 5–6.

[18] Alexander B. Murphy, "Regions as Social Constructs: The Gap between Theory and Practice," Progress in Human Geography 15 (1991): 30.

[19] See Harvey, Condition of Postmodernity.

[20] See Doreen Massey, Space, Place, and Gender (Minneapolis: University of Minnesota Press, 1994), chap. 7.

[21] So, for instance, the locational "fairness" of the Green Line, which demarcates the Israeli-occupied territories and thus, provisionally, an independent Palestinian state, hinges on questions of its relation to the geography of settlements (ancient and recent) and their agricultural lands, of groundwater movement, and of strategically significant heights of land). See David Newman and Ghazi Falah, "Small State Behaviour: On the Formation of a Palestinian State in the West Bank and Gaza Strip," Canadian Geogra-

One promising approach to the challenge of establishing a moral framework for accommodating the interests of a particular community within a set of potentially universalizable norms is offered by Seyla Benhabib's concept of the "concrete other."[22] In place of an ethic of the "generalized other," governed by the formal reciprocity of rights, obligations, and entitlements, she calls for one based on the complementary reciprocity of (intersubjective) sympathy, solidarity, care, and love. These moral sentiments draw upon humans' "inner nature," which has been repressed in the dominant traditions of Western moral philosophy and political theory. Both the communicative approach to ethical discernment that they require and their utopian thrust betray an intellectual debt to Habermas; but Benhabib develops her argument to identify the gendered blinkers of that "right-bearing, adult male," the "generalized other," and to clarify that only within a disciplined commitment to a wider community of justice and solidarity can plurality and difference be conscientiously embraced. In the contemporary era, even within societies such as those of the United States and Canada, that commitment cannot be assumed to rest on shared cultural values, so there is an unavoidable requirement for cross-cultural sensitivity, not least in the consideration of society-environment relations.[23] Any perceived novelty in the incorporation of concepts of space and place to inform discussion of environmental change and social justice is a measure of the erstwhile suppression of non-hegemonic voices, such as those of women, indigenous peoples, and minority cultures, whose worldviews have long embodied such knowledge.

A WORLD OF CORES AND PERIPHERIES

We have argued that the geographical diversity of the Earth's natural environment exists independently of the historical geography of the world's civilizations, yet the two are not totally unrelated. If we address questions of geographical scale and relative location more concretely, we can elaborate some of the ethical implications of space and

---

pher 39 (1995): 219–34, and David Newman, "Demarcating Israel's Boundaries: Creating the Map of Peace" (Philip E. Uren Memorial Lecture, Carleton University, Ottawa, March 1994).

[22] Seyla Benhabib, *Critique, Norm, and Utopia: A Study of the Foundations of Critical Theory* (New York: Columbia University Press, 1986), especially pp. 340–53.

[23] See Thomas McCarthy, "On the Pragmatics of Communicative Reason," in *Critical Theory*, ed. David Couzens Hoy and Thomas McCarthy (Oxford: Blackwell, 1994), especially pp. 90–93.

place. The dynamic processes whereby some societies/places become "core" regions and exercise power over space and particular "peripheries" pose intractable obstacles to the achievement of social justice.[24]

Although states come in a very wide variety of shapes, sizes, populations, resource potentials, and administrative competencies, the state continues to be the starting point for nearly all normative theorizing at the global scale. Each state is, at least in principle, sovereign over a particular tract of terrestrial space and its environmental resources. Yet the processes by which modern states and their antecedent societies came to achieve that control have uniformly been characterized by the exercise of aggressive or coercive power. Societies that came into being occupying *one* portion of the Earth have imposed themselves upon the peoples and territory of *another*. Tapping additional environmental resources has invariably been an explicit goal of such colonization. It clearly characterized the colonial empires of European states in the period between the late-fifteenth and mid-twentieth centuries, and equally the territorial expansion of Imperial Japan. But even the supposed nation-states of Europe owe their existence to prior conquest and forced assimilation (Normans over Saxons in England; the people of the Ile de France over those of the "langue d'oc," for example). Moreover, the varied legal systems of modern states, expressive primarily of the values of social and political elites, have been used to disallow the claims of subordinate groups (not least to environmental resources), to render illegitimate alternative institutional structures and knowledges (such as traditional rights of environmental usufruct), and to sanction by default the environmental destructiveness of the dominant industrial system.

At what point, and by what strategies of hegemonic discourse, have some of these histories become "naturalized," so that former resource depletion and past injustices done to subjugated peoples are no longer widely regarded as grounds for contemporary claims for redress? As an ontological category, "the state" is ethically compromised: the morally justifiable extent of any state's sovereignty over people, terrestrial space, and environmental resources invites critical and specific scrutiny. This involves questions about the concepts of place held by the world's inhabitants, in different regions, and at different geographical scales. At what level do attachments to place form, and to what extent do those shared attachments become constitutive of a communal identity? Western political theory has essentially restricted such

---

[24] See Charles Gore, *Regions in Question: Space, Development Theory, and Regional Policy* (London: Methuen, 1984).

identity to the level of the state, based on the assumption that its territorial boundaries precisely encompass a population of self-consciously distinct cultural identity—and, by extension, exclude those of different identity. This geographical scale of attachment to place, the state as community—as "nation"—was nurtured in the ideology of political elites in key European core states (England, France, Germany) and, in the "melting pot" variant, in the United States of America. Yet the congruence of state frontiers with the geography of self-conscious national communities is the exception rather than the rule. The great majority of the world's states are *not* "nation"-states. Both in the global core (e.g., Belgium, Switzerland, Canada) and throughout the formerly colonial periphery, state boundaries do not match the distribution of "peoples," communities of shared cultural identity, formed in part by their history of having shared a "place."

In many states, for a majority of inhabitants, the primacy of community identity is felt at more bounded scales of region or locality. There may be a very weak sense of "nationhood" (rhetorically dignifying "statehood"), especially where the institutional structure of the state has been imposed by "outsiders" of differing ethnicity.[25] Increasingly, "peoples" who have a strong sense of their own identity as a community with a shared history are asserting their "nationhood" over against the state (or states) in which their "place" lies. This may lead to efforts to bring about some territorial change or devolution of sovereignty that achieves a better fit between their territorial "home" and the recognition of that home within the international system of states.[26] The relatively peaceful pursuit of such measures in the arctic circumpolar region contrasts with the much more violent circumstances of Bosnia, Palestine, and some of the former Soviet republics in central Asia, for example.

What is plain is that within the institutions of the modern state system, the legitimacy of claims to self-determination, and with it the recovery of control over territory and environmental resources, is only selectively recognized. Territorial agglomerations of diverse populations constituted as discrete political entities by imperial powers (that is, states that are former colonies) are accorded legitimacy and rights

[25] And where, as in the Treaty of Berlin, which determined most of the political boundaries in Africa, state borders were imposed in ignorance and disregard of local biophysical parameters.

[26] See David B. Knight, "Identity and Territory: Geographical Perspectives on Nationalism and Regionalism," *Annals of the Association of American Geographers* 72 (1982): 514–31, and "People Together, Yet Apart: Rethinking Territory, Sovereignty, and Identities," in *Reordering the World: Geopolitical Perspectives on the 21st Century*, ed. G. J. Demko and W. B. Wood (Boulder, Colo.: Westview Press, 1994), pp. 71–86.

that have largely been denied to distinctive substate societies, both "indigenous peoples" and cultural or ethnic minorities.[27] These issues are well illustrated in the ambiguities of the contemporary aspirations to sovereignty on the part of Parti Quebecois (PQ) governments in the Canadian province of Quebec. The objective of the PQ is to secede from the Canadian federation with all the territory contained within the present provincial boundaries. This includes northern lands occupied almost exclusively by indigenous peoples, who insist on their right to remain within Canada, and over whose territory Quebec did not have jurisdiction at Confederation (1867). That the PQ's championing of the rights to self-determination of its French-speaking constituency vis-à-vis the Canadian state is not matched by its sympathy for the parallel claims of minority cultures vis-à-vis the provincial state has not gone unchallenged.

The now-fading global dominance of European-based societies reflects a complex history of resource exploitation and pattern of resource use that confounds any simple correlation of wealth and power with measures such as organic NPP at the world scale. But at the scale of continents and subregions thereof, it is more readily apparent that economic surplus derived from a more productive environment for agriculture has significantly shaped the geography of power.[28] Core-periphery or heartland-hinterland relationships are thus more than simply constructs of political economy: they rest on and embody environmental contrasts also. Today, heartlands are the centers of political and economic power and the home of the cultural/media elites—those who define "the issues." They are invariably situated in productive agricultural regions but are no longer economically dependent on resource-based industries (including agriculture). With respect to environmental resources, they are now centers of consumption rather than production. Peripheries exhibit the converse characteristics—politically and economically dependent, with heavier (in places all but complete) reliance on environmental resources as the basis of people's live-

[27] See David B. Knight, "The Dilemma of Nations in a Rigid State-Structured World," in *Pluralism and Political Geography: People, Territory, and State*, ed. Nurit Kliot and Stanley Waterman (London: Croom Helm, 1983), pp. 114–37, and "Territory and People or People and Territory: Thoughts on Post-Colonial Self-Determination," *International Political Science Review* 6 (1985): 248–72, and Will Kymlicka's chapter in this book.

[28] The location and disproportionate growth of London and Paris and their regions relative to the states within which they are set exemplifies this relationship, as does, at a different scale, the human geography of the Middle East and Southeast Asia. It is less marked in "settler" societies, where the accidental geography of first contact has shaped the political map, but it still holds for southern Ontario and southeastern Australia, for example.

lihood. And this polarization is reflected also, of course, at the global scale in the differentials of power and wealth that constitute the North-South divide.

To the extent that peripheries have been shaped by external institutions and forces, they have been created as places that relieve environmental stress in the core.[29] Their resources have been exploited to support a level of per-capita resource utilization in the heartland that could not be sustained locally; and/or they have been occupied by migrants from cores where pressures of population growth would have led to demographic involution without such relief.[30] (These are two sides of the same coin, but their dynamics are worth differentiating.) The process and purpose of migrant settlement or the establishment of expatriate enclaves, with or without the displacement of indigenous societies, has incontrovertibly contributed to the present configuration and vulnerabilities of hinterland economies and environments—regional characteristics that differ from those that would have been generated in the course of a more autonomous path of "development." There is, of course, a substantial debate over the net gains or losses that hinterland societies at the global, continental, or national scale have experienced through this process. The bidirectional pattern of core-periphery resource flows that characterized relations between European imperial nations and their overseas colonies, for instance, differs substantially from that associated with the pattern of transfer payments established among the regions of Canada by the postwar federal welfare state.[31] But even in the latter context, argument arises whether the mechanisms through which core societies achieved their ascendancy impose *obligations*, as distinct from the recognition of political expediency, on them to compensate for the induced differentials of power, prosperity, and environmental vulnerability. Given the counterfactual nature of inferred paths of development in peripheries had they not been incorporated by core institutions, together with by no means unprincipled critiques, in the case of many post-independence

[29] Among recent work to give analytical substance to this claim is William E. Rees and Mathis Wackernagel, "Ecological Footprints and Appropriated Carrying Capacity: Measuring the Natural Capital Requirements of the Human Economy," in *Investing in Natural Capital: The Ecological Economics Approach to Sustainability*, ed. Ann Mari Jansson et al. (Washington, D.C.: Island Press, 1994), pp. 362–90. The footprint of the Netherlands, as an example of an economy in the global core, is estimated to occupy an area fourteen times larger than the territory of the state.

[30] Notably in the overseas migration that relieved demographic pressure in nineteenth-century Europe.

[31] See A. J. H. Latham, *International Economy and the Underdeveloped World, 1865–1914* (London: Croom Helm, 1978); Economic Council of Canada, *Living Together: A Study of Regional Disparities* (Ottawa: Supply and Services Canada, 1978).

Third World states for instance, of autonomous policy choices governing areas as diverse as exchange rates and natalism, it has been relatively easy for core elites to deflect the moral claims of the periphery.
This dynamic was very evident in the United Nations Conference on
Environment and Development (UNCED) at Rio de Janeiro in 1992.[32]

Nevertheless, in many parts of the world there is a new sense of
obligation to redress what are now acknowledged as injustices inherent in the historical geography of core-periphery interaction. Moves
such as those in Canada, not only to restore self-government, and with
it managerial control over terrestrial resources, to indigenous peoples
or communities, but also to make financial provisions to give substance to that renewed independence (in a very different world from
that in which it was forfeited), reflect this changed atmosphere.[33] Majority public opinion in the Euro-Canadian state has agreed to compensate, however inadequately in an ultimate sense, for past practice.
Comparable action has been taken by white settler societies in New
Zealand and Australia.[34] A deeper and contextually very different challenge has been taken up by the white citizens of South Africa. The
uncertain outcome of that country's current social transformation will
entail devising relationships of justice between people, land, and natural resources with concrete territorial expression.[35]

SPACE, PLACE(S), AND THE "CONCRETE OTHER"

If the social and political polarization that separates cores and peripheries and forms a major barrier to the achievement of just solutions to
the challenges of environmental change is to be bridged, much more
effort needs to be put into comprehending the concreteness of "the
other." As a consequence of the embeddedness of people and their
consciousness in the places where they live, protagonists in conflicts
over the use or abuse of Nature, across the full range of geographical

[32] S. B. Hecht and A. Cockburn, "Realpolitik, Reality, and Rhetoric in Rio," *Society
and Space* 10 (1992): 367–75.

[33] David F. Pelly, "Dawn of Nunavut," *Canadian Geographic* 113 (March/April
1993): 20–29.

[34] See, for instance, W. H. Oliver, *Claims to the Waitangi Tribunal* (Wellington, New
Zealand: Department of Justice, 1992), and David Mercer, "*Terra nullius*: Aboriginal
Sovereignty and Land Rights in Australia: The Debate Continues," *Political Geography*
12 (1993): 299–318. Antony Turner, "Mabo: The Biggest Question Facing Australian
Mining," *CRS Perspectives* (Kingston, Ontario), no. 47 (1993): 3–12, interprets a key
Australian High Court decision from the perspective of the resource-based industry
most affected.

[35] See Charles Villa-Vicencio, *A Theology of Reconstruction: Nation-Building and
Human Rights* (Cambridge: Cambridge University Press, 1992).

scales from the global to the local, are obliged to reckon seriously with the full humanity and specific terrestrial context of their opponents: to come to terms not only with a "we" and a "they," but to acknowledge a "you." As Thomas McCarthy elaborates, the need to adopt a consistently global perspective, and thereby uphold a commitment to the putatively normative reason identified with the "generalized other," requires combining "a strong historical sense of who we are and who we want to be with an equally full-bodied comparative sense . . . shaped by [dialogue with] those whose full humanity has been somehow denied. . . . [For only in such a context might] the boundless claims of reason . . . finally prove themselves to be more than Eurocentric illusions."[36] Justice will only be achieved on the basis of interaction between parties who encounter the concrete identity and place of "the other."

As one consequence, discourse in which a global "community" is challenged to respond to "global" environmental change becomes problematic. Simply to trace the slippery semantics of "we" on page 1 of the Bruntland Commission Report is enough to establish that the primary voice is that of those raised in the rich industrialized countries.[37] While there is an obvious sense in which contemporary environmental problems affect an integrated global biosphere, and economic disruptions reverberate throughout a relatively open global economy, there is plenty of room to question the degree to which rich and poor nations face a "Common Future." The particular hegemonic perspective that so defines it is contested by the "Southern" response, For Earth's Sake, which pointedly eschews talking about "global" environmental change because of the term's cooptation in "Northern" discourse to distance the industrialized world from the "local" environmental problems of the less developed countries.[38] The emotive icon of the Earth photographed from space has been used in the North to foster the notion of a single global community inhabiting "Spaceship Earth." But a much more accurate image of humankind, deeply divided between a few rich and many poor, is the champagne-glass graphic of the world's population as structured by wealth distribution that encapsulated the message of the United Nations' 1992 Human Development Report.[39]

[36] McCarthy, "On the Pragmatics of Communicative Reason," pp. 92–93.
[37] The World Commission on Environment and Development, Our Common Future (Oxford: Oxford University Press, 1987).
[38] The Commission on Developing Countries and Global Change, For Earth's Sake (Ottawa: International Development Research Centre, 1992).
[39] United Nations Development Program, Human Development Report 1992 (New

It bears repeating that the prevailing patterns of core-periphery/
North-South/national heartland-hinterland disparities that characterize the world's human populations arise out of society-environment
relations that are neither entirely "natural" nor entirely sociopolitical
artifacts. Nevertheless, the responsibility of core societies and institutions for shaping these patterns is increasingly difficult to "naturalize"
through a hegemonic discourse. The historical geography of their creation, and of how each regionalized community and economy of resource use has been constituted, is increasingly recognized as a salient
ethical dimension of the contemporary challenges of environmental
change. And because "concretely other" places are partly constitutive
of "concretely other" communities, an environmental ethic cannot be
divorced from a social ethic. "Doing right" by the forest (whether in
the Amazon or British Columbia), or the soil and aquifers of the Great
Plains, or even the biophysical environment of urban settlements from
Athens to Bangkok, must include "doing right" by those people whose
immediate environment it is. The legitimacy of this argument is today
widely accepted with respect to indigenous peoples, for they and their
land are accepted by most outsiders as constituting a "community,"
and their history is acknowledged as one of being "more sinned
against than sinning." But the moral claim of historically immigrant
populations whose past colonization of space and occupancy of place
is today contested rests on weaker foundations. The settler societies of
the Americas, Australasia, and parts of Africa created very different
forms of environmental resource–based communities than those of the
indigenous forest/rural/nomadic societies that gradually evolved in
most of the world (including preindustrial Europe) prior to the nineteenth century. The ethical issues raised by the European land grab in
Kenya, Algeria, or Zimbabwe were perhaps relatively clear-cut. A
longer history of occupancy and environmental stewardship has not
been sufficient to naturalize fully the presence of the Afrikaaners in
South Africa. Is the moral basis of the dominant society in Canada or
the United States any more secure, or are the problems that would
arise in radically reviewing it simply less tractable?[40]

At the continental or (large) national scale, the vexed questions of
reconciling environmental and social ethics are most obvious, but not

York: Oxford University Press, 1992). The highest quintile of the world's population
controls 82.7 percent of global income; the next, 11.7 percent; and the remaining three
combined, 5.6 percent.

   [40] Note also the difficulties of agrarian restructuring following the collapse of the com-
munist regimes in Russia and Eastern Europe associated with the revival of private
claims to state-confiscated land.

necessarily most salient, in the periphery. Why is this? In the first instance because it is on the peripheries of dominant societies that subordinated peoples are proportionately most numerous and best able to challenge the hegemony of core institutions. Second, by choice (as in the arctic) or through forced relocation (as in the American West), the periphery is the "place" of indigenous peoples, the setting in which conflict between their environmental values and practices and those of an industrialized economy of "resource" exploitation is likely to be most sharply defined.[41] But equally, many parts of resource-rich peripheries have become, over a number of generations, the "place" of non-native communities, many of whose members have ties to and concerns for their biospheric environment and a livelihood derived from it. They represent "concrete others" whose interests have an equal claim to moral standing (a point we return to below). A third reason for the location of disputes involving a clash of environmental values in the periphery, however, is that resource exploitation or environmental degradation of varying degrees of severity is usually readily visible (from the air if not from the ground) and in many contexts sufficiently site-specific to be the target of effective campaigns of opposition orchestrated primarily, though not exclusively, by members of *core* communities.[42] By means of legal powers or political influence (involving core-dominated institutions), or of direct action (often made more effective by core-based media attention), external voices claiming the moral high ground seek to impose their conception of ethical environmental practice on communities in distant places.[43]

Yet respect for the concreteness of "the other" demands that one question whether the resource exploitation and environmental degra-

[41] This is not to deny that *within* the arctic, Canadian Inuit have also been subjected to involuntary relocation. See Frank Tester and Peter Kulchyski, *Tammarnlit (Mistakes)—Inuit Relocation in the Eastern Arctic, 1939–1963* (Vancouver: University of British Columbia Press, 1994), and Ronald Wright, *Stolen Continents: The New World through Indian Eyes since 1492* (Toronto: Viking, 1992).

[42] This should not be interpreted to imply that the environmental attitudes and behaviors of activists resident in core regions are indistinguishable from those of their less committed neighbors. But the activists remain distant (in more ways than one) from the communities their actions impact, and their own livelihood is rarely put at risk by their actions. On contrasting environmental knowledge and values between indigenous and nonindigenous people in the same region, see J. J. Shute and D. B. Knight, "Obtaining an Understanding of Environmental Knowledge," *The Canadian Geographer* 39 (1995): 101–11.

[43] The actions of West European environmentalists toward the fur-trapping of native communities in the Canadian Arctic provide telling examples. See Finn Lynge, *Arctic Wars, Animal Rights, Endangered Peoples* (Hanover, N.H.: University Press of New England, 1992), and George Wenzel, *Animal Rights, Human Rights: Ecology, Economy, and Ideology in the Canadian Arctic* (Toronto: University of Toronto Press, 1991).

dation occasioned by consumers in the core is any more ethically defensible than that of primary producers in the periphery.[44] The major instrumental difference, perhaps, is that the former is much less localized and immediately visible, and hence less vulnerable to effective opposition, whether from within or external to the core. It is easier, and usually more effective, for core-based environmentalists to block a logging road or to lobby for the refusal of a specific mine proposal in the periphery than to attack automobile pollution by cordoning off the freeways of a core metropolitan area. (Contrast the progress in Washington in the early days of the Clinton administration of legislating spotted owl protection in resource-dependent hinterland regions of the Pacific Northwest versus that of imposing a national energy tax that would hit home in suburbia.)[45] These heartland-hinterland contrasts in social power and the vulnerability of community livelihoods to interventions by people from another "place" challenge the moral integrity of society's adjustment to more sustainable patterns of resource use. A just solution to conflicts of value concerning patterns of environmental resource appropriation will oblige heartland elites to reckon with the concreteness of communities in the periphery, and to design approaches to environmental policy that are evenhanded in their imposition of livelihood adjustments on core and periphery alike.

But to "reckon with" does not imply uncritical acceptance of local practice. Environmental impacts of human action are rarely confined to the immediate area, and their consequences affect various communities of interest. What are the relevant communities in terms of their composition and geographical location? And whose or which interests deserve to be privileged? The ethics of acting as a citizen of the global "biospheric community" are not as straightforward as many activists would suggest. Anyone can claim on these grounds (though note the socioeconomic profile of those who actually do, and the "places" that they come from) that an often distant regional environment such as Amazonia or the British Columbia rain forest is part of a "global" biophysical endowment, too valuable for its fate to be left up to the priorities of the relevant state or local people to determine. Indeed, a radical biocentric stance would argue unambiguously that the natural environment has to be "saved" at any cost to the human communities living within it. In a North/South context, the pursuit of such an agenda is readily cast as an environmentalist version of neoimperialism. But within "white settler states" such as Australia, Canada, and

---

[44] This is, of course, precisely the question posed by the South in the global context.
[45] "Imitating Nature," *The Economist*, 6 November 1993, 26–27; "Tax Notes from the Bayou," *The Economist*, 5 June 1993, 25–26.

the United States, the legitimacy of the hinterland economy of industrialized resource extraction is frequently sufficiently ambiguous to strengthen arguments in favor of the imposition of bioethical imperatives from a distant core.

Yet it is suggestive that "the people simply shouldn't be there" is a response common to biocentric environmentalists in a context such as this and to neoclassical economists faced with regions of persistently lagging economies. In both cases the policy prescription embodies an idealized norm for a "generalized other" that dismisses the concrete history and bonds of community of the populations affected as irrelevant to moral or rational evaluation. The Canadian province of Newfoundland, which has long experienced the highest unemployment rates in Canada, provides an illustration. The standard policy response espoused by mainstream economists has been that residents should migrate to regions of greater economic opportunity, which is fair enough up to a point and is, of course, what young males have long done. But the age/sex selectivity of that migration is precisely one of the issues that render the neoclassical "solution" less than adequate. Now, with the dramatic collapse of the cod fishery on which traditional Newfoundland society has been built (and on which it has been made dangerously overdependent for a generation by government policies that inhibited social change and economic diversification), the ecological argument that the island is overpopulated has also been voiced. Responsibility for the collapse of the fish stocks is disputed and certainly complex. What is clear is that "traditional environmental knowledge" within the local community, which warned that the fishery, as recently practiced, was becoming unsustainable, was consistently discounted by the core-based scientific and business elites until too late in the day.[46]

Certainly, critical dialogue involving both core- and periphery-based communities is likely to identify needs to rectify, desist from, or compensate for past practice that has resulted in environmental degradation. The moral and practical force of new forms of society-environment interaction, however, rests on those whose lives are embedded in existing practices having the right to be heard and involved in the transition to a new order. The Government of British Columbia's attempt to do this for logging in Clayoquot Sound has shown the difficulties of reconciling the local with external (biospheric?) community

[46] See J. A. Agnew, "Devaluing Place: 'People Prosperity versus Place Prosperity' and Regional Planning," *Society and Space* 2 (1984): 35–45, and Raymond A. Rogers, *The Oceans Are Emptying: Fish Wars and Sustainability* (Montreal: Black Rose, 1995).

interests.[47] One measure of the adequacy of a social ethic is its capacity to protect the rights of the poor or marginalized—those living on the "sharp end" of the situation it addresses. But who are these in such a case? Are they represented by the distant environmentalist, living in another "place" but driven, without material interest, by the values of a biospheric ethic to oppose the hegemonic project of "development" (for example, the Ontario schoolteacher on the Vancouver Island roadblock); or are they represented by the lumberjack, a breadwinner tied to a place-bound community and the economic opportunities it affords, vulnerable to shifts in global markets or the corporate priorities of a distant employer, and now also to an external environmental movement that agitates to disallow a livelihood "at home"? If humanity consists of communities of people within their biophysical setting, in their "place," then the identification of value in a regional environment must be accompanied by a recognition of value in the lives of those who inhabit it.

The call to prioritize protection of the environment over its exploitation, even if "sustainably," cannot be discounted just because it comes from a distant place. All sorts of questions about whether the same resource development ends can be attained less destructively, or whether an alternative economic base offers a more sustainable future (tourism is a deceptive favorite), may well be appropriate. But the fundamental challenge facing most who argue for the application of a global biospheric ethic to a particular, and often distant, hinterland environment is to encounter the "concrete others" whose lives are bound up in that place. Concern for their welfare is currently the exception among core-based environmental movements.[48]

OUR world, the world of humankind and its biospheric ecumene, is very diverse. It is full of peoples, *places*, and environments. Change is constant, within the natural and human worlds, and in the complex interactions between them. What is new in the contemporary era is the recognition of how deeply and potentially destructively "modern" societies are implicated in the processes of biophysical change. The flow of our discussion has been leading us, moreover, to argue against the

[47] Province of British Columbia, *Clayoquot Sound Land Use Decision: Background Report* (Victoria, B.C.: Commission on Resources and the Environment, 1993).

[48] See Dorceta E. Taylor, "Women of Color, Environmental Justice, and Ecofeminism" (School of Natural Resources and Environment, University of Michigan, 1994). The disproportionate concentration of toxic waste sites in localities occupied by minority and low-income populations within the United States has been extensively documented. See Robert Bullard, *Dumping on Dixie: Race, Class, and Environmental Quality* (Boulder, Colo.: Westview Press, 1990).

separation of "environmental" and "social" ethics, on account of the embeddedness of human societies in "places" where a shared history and experience of encounter with the natural environment, and its cultural transformations, has created a particular sense of identity-in-place. This self-understanding requires ethical evaluation, especially of its claims (implicit or otherwise) that conflict with those of other people in other places, and of communities defined at other geographical scales.[49] The pursuit of social justice, however, cannot avoid the legacy of how these claims have evolved or of how they have been shaped by differentials of power among peoples to transform or appropriate environmental resources. Giving substance to the vision of a global community with a common future requires negotiating an equitable accommodation to new pressures and new constraints in each particular place and community.

The challenge of holding the two scales of resolution in tension should not be underestimated. Given the interdependencies of the Earth's biophysical systems and of the contemporary global economy, the ethics of "community" building and environmental practice in one part of the world cannot ignore the impact that choices impose on communities and environments elsewhere. General principles that will promote the realization of social justice in a time of accelerating environmental change are vitally necessary, but they will serve this cause only if framed to accommodate the concrete variety of "places" that comprise the "space" of the global ecumene and the "moral particularities" of the lives that are lived in it.

[49] Core elites need to acknowledge, inter alia, that "peripheral" communities perceive themselves as being "in the center" in terms of their own worldview, and hence that policy determined in the core needs to accommodate this peripheral self-understanding. See D. B. Knight, "Geographical Considerations in a World of States," in *States in a Changing World*, ed. R. H. Jackson and A. James (Oxford: Clarendon Press, 1993), pp. 26–45, and Neil Smith, "Homeless/Global: Scaling Places," in *Mapping the Futures: Local Cultures, Global Change*, ed. Jon Bird et al. (London: Routledge, 1993), pp. 87–119.

5

The Normative Structure
of International Society

CHRISTIAN REUS-SMIT

WHEN we consider the ethical status of international environmental accords we need to address two distinct but interconnected levels of justice. Henry Shue has usefully characterized these as *background justice* and *internal justice*.[1] International agreements are negotiated within preexisting social, economic, and political contexts, and we speak of background justice when we assess the justice of these circumstances. Internal justice refers to the justice of the accords themselves, whether the terms of the agreements can be considered fair or unfair, equitable or inequitable, empowering or oppressive. The two levels of justice are obviously connected. If background conditions are unjust, leaving some actors disadvantaged, then it is difficult to imagine the resulting agreements being internally just. As these internally unjust agreements are institutionalized, they become part of the background circumstances in which new accords are formulated.

Some theories of justice are better suited to questions of background justice, whereas others are more appropriate to issues of internal justice. The relative appropriateness of a given theory is largely determined by its underlying ontological assumptions. Every theory of international justice presupposes a particular account of the nature and

The author thanks Amy Gurowitz, Richard Price, Henry Shue, Sarah Soule, and the editors, whose insights greatly influenced and sharpened the ideas explored here. The research and writing of this chapter were financially supported by an SSRC-MacArthur Foundation Fellowship in Peace and Security in a Changing World.

[1] Henry Shue, "The Unavoidability of Justice," in *The International Politics of the Environment*, ed. Andrew Hurrell and Benedict Kingsbury (Oxford: Clarendon, 1992), p. 387.

dynamics of international relations.[2] Whether explicitly enunciated and developed or buried within layers of philosophical argument, these accounts perform two crucial functions. First, they clarify the social arena under moral scrutiny. Second, they define the realm of ethical possibility, that is, they inform the theorist's conclusions about the potential for, and the limits of, moral action and change. For instance, Michael Walzer's controversial assumptions about national political communities uphold his claims about the moral standing of states and the impossibility of international distributive justice, whereas Charles Beitz's assumptions about international interdependence support his claims about the desirability and possibility of distributive reforms beyond borders.[3]

Here I am concerned with two contrasting theories of international justice—*distributive theory* and *critical theory*—and their relative suitability for assessing the justice or injustice of international environmental accords. Until now most serious work on international environmental justice (as opposed to environmental ethics in general) has been done from a distributive standpoint. According to this perspective, of which Shue's work is exemplary, international environmental accords are assessed according to whether or not they provide a fair distribution of benefits and burdens among rich and poor countries, which usually means a fair distribution of the costs of adjustment necessitated by global environmental protection.[4] This approach has obvious relevance, given the prominence of distributive issues in current international environmental negotiations; however, I contend that distributive theories of justice are best suited to questions of internal justice, and their value declines markedly when applied to problems of background justice.

Distributive theories of international justice are founded on an *individualist* account of international relations that does not adequately capture the background conditions in which international environmental accords are negotiated. Stated briefly, the norms, rules, and principles that constitute international society and define the institutional context in which environmental negotiations take place are not

[2] For an elaboration of this line of argument, see Carol Gould, *Rethinking Democracy: Freedom and Social Cooperation in Politics, Economy, and Society* (Cambridge: Cambridge University Press, 1990), p. 91.

[3] See the essays by Walzer and Beitz in *International Ethics*, ed. Charles Beitz et al. (Princeton: Princeton University Press, 1985). Also see Charles Beitz, *Political Theory and International Relations* (Princeton: Princeton University Press, 1979).

[4] See Henry Shue's chapter in this book. Also see his articles "The Unavoidability of Justice" and "Subsistence Emissions and Luxury Emissions," *Law and Policy* 15 (January 1993): 39–59.

easily understood in distributional terms. By contrast, a critical-theoretical conception of international justice builds on a *societal* account of international relations, one that emphasizes the intersubjective, unquantifiable, and nondistributable nature of institutional norms, rules, and principles and thus sustains a quite different conception of international justice. Background justice is no longer assessed in distributional terms, but rather in terms of actors' abilities to participate freely in the moral discourse that generates new social institutions. Although the distribution of material goods clearly influences such participation, a critical theory of justice focuses on ideological constraints, on how normative structures empower some actors and marginalize others, and on how prevailing moral frameworks license certain moral claims and silence others.

There are three questions that from the standpoint of a critical-theoretical conception of justice, are central to our assessment of the background circumstances that frame international environmental negotiations. First, which background institutions of international society have the greatest implications for the justice of environmental accords? To answer this question one must focus on the deep normative structure of international society, on the very raison d'être of the modern sovereign state, or what I have termed *the moral purpose of the state*.[5] Most international relations scholars treat sovereignty as the primary value of international society, its most basic "organizing principle." I suggest that sovereignty is best thought of as a derivative value. Throughout history justifications for the internal authority and external autonomy of the state have rested on prior claims about the moral purpose of centralized, autonomous political organizations, claims that have varied from one system of sovereign states to another. A distinctive moral purpose has provided the foundational norms of the modern system of states, norms that endow certain actors with legitimacy while denying others standing, norms that sanction certain sorts of moral discourse over others.

Second, how does this deep normative structure influence actors' abilities to participate in the processes of environmental norm formation? I contend that the moral purpose of the modern state, the rationalistic pursuit of economic and social progress, structures international environmental negotiations in a fundamentally unjust manner. The core of modern states' raison d'être is a rationale that derives from a particular set of historically contingent ideas about human agents, about their relationship to their society at large, and about the

---

[5] Christian Reus-Smit, *The Moral Purpose of the State: Social Identity, Legitimate Action, and the Construction of International Institutions* (Princeton: Princeton University Press, in press).

role of political organization. States that assume this mantle, in substance or in rhetoric, constitute the legitimate actors of modern international society. Furthermore, moral claims that resonate with this ideal of the modern state, whether they come from state or nonstate actors, have particular salience in the formation of regime norms. These observations help us understand the patterns of inclusion and exclusion that now characterize international environmental negotiations. To illustrate these patterns, I briefly examine the experiences of three differently positioned sets of actors who have sought to influence the formation of environmental norms. The principal boundary lies between those actors who can define their moral claims in modernist terms and those whose position or value-orientation deny them this possibility.

Third, what are the prospects for meaningful international change? I contend that we must redefine the moral purpose of the modern state before we can develop truly just and effective international environmental institutions. In assessing the likelihood of such a transformation, I offer some preliminary observations about the general conditions under which deep normative structures of systems of states change, about whether or not such conditions presently exist, and about where the catalyst for change is likely to originate. It just may be possible to redefine the moral purpose of the modern state under present historical conditions, but the impetus for change will have to come from below.

FROM DISTRIBUTIVE JUSTICE TO CRITICAL-THEORETICAL JUSTICE

Individualist Ontology and Distributive Justice

Distributive theories of justice, in both their domestic and international manifestations, assume an individualist conception of human agency and social interaction. This view rests on three general assumptions: individuals are atomistic, rational, and possessive; they are autonomous sources of their own conceptions of the good; and they engage in social relations primarily to obtain benefits that autonomy and self-help cannot provide. These benefits include the provision of a secure and stable environment in which to pursue their individual life plans, and the exchange and collective production of the resources needed to realize such plans.[6]

The link between this individualist ontology and the distributive

[6] Charles Taylor calls these views "atomistic." See "The Nature and Scope of Distributive Justice," in Philosophy and the Human Sciences: Philosophical Papers 2 (Cambridge: Cambridge University Press, 1985), p. 292.

theory of justice is usefully illustrated by considering David Hume's conception of justice, one of the earliest modern formulations of distributive ethics. Outside society, Hume argues, individuals lack secure access to, and control over, the resources needed to pursue their particular life plans. Guaranteeing stable possession is thus the principal reason why individuals enter society. "No one can doubt," Hume argues, "that the convention for the distinction of property, and for the stability of possession, is of all circumstances the most necessary to the establishment of society." Stable possession, however, requires principles of individual entitlement, a way of determining right ownership. "Some method must be shewn," Hume writes, "by which we may distinguish what particular goods are to be assign'd to each particular person, while the rest of mankind are excluded from their possession and enjoyment." These principles of property entitlement constitute the essence of justice. "Our property," Hume concludes, "is nothing but those goods, whose constant possession is established by the laws of society; that is, by the laws of justice."[7] Thus in a world of atomistic individuals, each pursuing his or her own conceptions of the good, each seeking resources within a finite environment, justice entails the right distribution of goods.[8]

The individualist ontology exerts a profound influence on theories of international cooperation, particularly rationalist theories of international regimes. Arthur Stein asserts that the "same forces that lead individuals to bind themselves together to escape the state of nature also lead states to coordinate their action, even to collaborate with one another." Having drawn this analogy, rationalists go on to embrace the three foundational assumptions of the individualist ontology. States, like individuals, are considered atomistic, rational, and possessive. Their interests are assumed to be "autonomously calculated" and thus exogenous to social interaction.[9] And they are said to engage in

[7] David Hume, *A Treatise on Human Nature* (Oxford: Clarendon Press, 1989), pp. 488, 491, 502. This interpretation of Hume's account of justice is not incompatible with other readings that emphasize the importance of moral sentiments, especially sympathy, in Humean ethics. It is important to stress, however, that for Hume justice is an artificial virtue, the product of self-interest in the first instance and sentiment only later (p. 533).

[8] The reasoning that connects the individualist ontology with the idea of justice as right distribution is common to all theories of distributive justice. For instance, note the similar logics underlying the otherwise distinctive positions elaborated by John Rawls and Michael Walzer. See Rawls's *A Theory of Justice* (Oxford: Oxford University Press, 1971), p. 4, and Walzer's *Spheres of Justice: A Defence of Pluralism and Equality* (New York: Basic Books, 1983), p. 3.

[9] Arthur Stein, "Coordination and Collaboration: Regimes in an Anarchic World," in *International Regimes*, ed. Stephen Krasner (Ithaca: Cornell University Press, 1983), p. 132.

cooperative social relations and create regimes to achieve goals that autonomy and self-help cannot provide.[10] International institutions, according to this perspective, are both the product of cooperation among egoists and a necessary precondition for the negotiation of future agreements.

The rationalist approach to international cooperation and regimes, along with its underlying individualist ontology, provides the framework for most analyses of international environmental politics.[11] In their introduction to *The International Politics of the Environment*, Andrew Hurrell and Benedict Kingsbury affirm the image of international society as "a practical association" of independent, self-interested actors. The "structure of both the international political and legal systems," they claim, "continues to rest heavily upon the independence and autonomy of separate sovereign states and the pluralism which this entails." It follows, therefore, that states will only "agree to, and commonly comply with, international environmental agreements because it is in their interests to do so." When agreements are achieved, Hurrell and Kingsbury argue, they will "reflect the shared interest in the collective management of a particular environmental problem . . . and embody a stable outcome to the underlying bargaining process between states over the distribution of the costs of collective action and other politically divisive issues." The challenge animating this and other studies of international environmental politics is to determine the conditions under which atomistic, self-interested states can cooperate to overcome the collective action problem posed by the environmental crisis.[12]

Just as most theories of international environmental cooperation are built on the individualist ontology, so too are most studies of international environmental justice. Here the question is not the conditions under which egoistic states cooperate to achieve international environmental protection, but rather the justice or injustice of how these actors distribute the benefits and burdens of their cooperation. Working from the individualist ontology transposed into the international

[10] Robert Keohane, *After Hegemony: Cooperation and Discord in the World Political Economy* (Princeton: Princeton University Press, 1984), p. 88.

[11] This is especially true of two recent edited collections on the subject. See Hurrell and Kingsbury, *The International Politics of the Environment*, and Peter M. Haas, Robert Keohane, and Marc Levy, eds., *Institutions for the Earth: Sources of Effective International Environmental Protection* (Cambridge: MIT Press, 1993). Rationalist assumptions also pervade Oran Young's work on international environmental politics. See *International Cooperation: Building Regimes for Natural Resources and the Environment* (Ithaca: Cornell University Press, 1989).

[12] Hurrell and Kingsbury, *The International Politics of the Environment*, pp. 6, 23, 1.

arena, justice is thus understood as right distribution. In his contribution to this book, Shue provides one of the best, and most thoroughly developed, applications of this approach. He begins by identifying four questions that together provide a general framework for assessing the moral status of international environmental accords dealing with global warming:

1. What is a fair allocation of the costs of preventing the global warming that is still avoidable?
2. What is a fair allocation of the costs of coping with the social consequences of the global warming that will not in fact be avoided?
3. What background allocation of wealth would allow international bargaining—about issues like (1) and (2)—to be a fair process?
4. What is a fair allocation of emissions of greenhouse gases (a) over the long-term and (b) during the transition to the long-term allocation?[13]

The imprint of the individualist ontology on Shue's work is apparent in two important respects. First, the questions outlined above are all allocative; they concern the right distribution of costs and benefits between individual actors. Second, his argument about what constitutes right distribution rests on the moral priority of certain individual rights. Poor countries, he argues, should not be asked to sacrifice their economic development in the name of global environmental protection because their right to economic development has moral priority. Building on his arguments in *Basic Rights*, Shue posits a hierarchy of human rights, grounded in a hierarchy of human needs.[14] Economic development, he argues, is directly related to the very survival of the world's poorest people and thus has moral priority over the luxury interests of the rich. Justice, therefore, "does not permit that poor nations be told to sell *their* blankets in order that rich nations may keep *their* jewellery."[15]

Distributive Justice and the Problem of Institutions

The inequitable structure of the international economy, combined with the inevitable costs of adjustment to environmental change and

---

[13] Shue, "Environmental Change and the Varieties of Justice," p. 17 (herein).
[14] Henry Shue, *Basic Rights: Subsistence, Affluence, and U.S. Foreign Policy* (Princeton: Princeton University Press, 1980).
[15] Henry Shue, "Unavoidability of Justice," p. 397.

protection, mean that distributive issues will remain central to international environmental negotiations. The Multilateral Fund established under the London revisions to the Montreal Protocol and the Financial Mechanism of the United Nations Framework Convention on Climate Change are but two examples of the impact such issues have already had on international environmental agreements. Given the preeminence of these issues, questions of distributive justice must obviously play an integral part in our assessments of the ethical status of environmental accords. And as Shue correctly points out, these issues arise at two levels: at the level of internal justice (the justice of the accords themselves), and at the level of background justice (the justice of the circumstances in which agreements are negotiated).[16]

This having been said, it would be wrong to assume that all questions of international environmental justice are of a distributive nature. This is especially true for background justice. My central claim is that states negotiate environmental accords within a preexisting complex of social institutions and that these institutions represent the fundamental background circumstances of justice. Institutions are complexes of norms, rules, and principles that "define the meaning and identity of the individual and the patterns of appropriate economic, political, and cultural activity engaged in by those individuals."[17] They occur simultaneously at four levels of international society: issue-specific regimes constitute the fourth and most superficial level; the fundamental institutions of international law and diplomacy represent the third, more enduring level; the organizing principle of sovereignty occupies the second level, and the underlying moral purpose of the modern state, its raison d'être, comprises the most primary, foundational level. The justice or injustice of these institutions, and institutions in general, is difficult to assess from a distributive standpoint.

Theories of distributive justice depend on the ability to divide goods according to principles of right entitlement. This process of division involves the weighing and valuing of goods and their allocation in just shares. When Hume and Locke first defined justice in distributive terms this process did not pose an insurmountable problem. Considerations of justice were then restricted to the morally proper allocation of material goods, that is, with property rights. Such goods were relatively easily divided and distributed. With time, however, the logic of

---

[16] Ibid., p. 387.
[17] John W. Meyer, John Boli, and George M. Thomas, "Ontology and Rationalization in the Western Cultural Account," in *Institutional Structure: Constituting State, Society, and the Individual*, ed. George M. Thomas, John W. Meyer, Francesco Ramirez, and John Boli (London: Sage, 1989), p. 12.

distribution was extended to cover the entire spectrum of social goods. For instance, William Galston argues that "issues of justice involve not only the distribution of property or income, but also such non-material goods as productive tasks, opportunities for development, citizenship, authority, honor, and so on."[18] Broadening the scope of distributive justice in this way required a fundamental reconception of society: "Human society is," Walzer claims, "a distributive community."[19] On close inspection, however, this new agenda and its associated image of society stretches distributive theory beyond comfortable limits.

In her excellent critique of distributive theories of justice, Iris Marion Young argues that the logic of distribution cannot be applied to the institutional context that determines patterns of distribution. This context includes "any structures or practices, the rules and norms that guide them, and the language and symbols that mediate social interactions within them." The norms, rules, and principles that comprise this environment cannot be weighed, valued, and allocated in the manner assumed and required by the distributive paradigm. The paradigm only comprehends the nature, occupancy, and power of decision-making positions, but cannot assess the rules and procedures according to which decisions are made. It also fails to comprehend nondistributive aspects of the division of labor. Although it addresses the allocation of jobs, it has little to say about the definition of occupations and the social status they attract. Finally, the distributive paradigm cannot assess the justice or injustice of cultural forms, by which Young means "the symbols, images, meanings, habitual comportments, stories, and so on through which people express their experience and communicate with one another." Such intersubjective meanings, she observes, "significantly affect the social standing of persons and their opportunities."[20]

Young's argument about the unquantifiable and nondistributable nature of institutional norms, rules, and principles dovetails nicely with the recent critique of positivist approaches to the study of international regimes. Positivist forms of explanation assume a model of causality that is essentially incompatible with how institutional norms affect human identities, interests, and behavior. As Friedrich Kratochwil argues, the positivist model understands "causality as force," which assumes a relation of power or compulsion between two dis-

[18] William Galston, *Justice and the Human Good* (Chicago: University of Chicago Press, 1980), p. 6.
[19] Walzer, *Spheres of Justice*, p. 3.
[20] Iris Marion Young, *Justice and the Politics of Difference* (Princeton: Princeton University Press, 1990), pp. 22–23.

crete and atomistic entities.[21] Norms, however, do not influence behavior in this way; they are intersubjective meanings that guide, inspire, and justify action. As such they violate the two principal requirements of a positivist model of explanation: because they are intersubjective meanings they cannot be isolated or measured, and because their causal impact is "constitutive," they are not "forces." Kratochwil and John Ruggie conclude, therefore, that a serious disjuncture exists between the positivist epistemology guiding the study of international institutions and the social ontology of international regimes.[22]

It is the positivist logic of distributive theories of justice—their inherent need to weigh, value, and apportion—that Young argues cannot accommodate the intersubjective dimensions of social institutions. And it follows, if Kratochwil and Ruggie are correct, that distributive theories of international justice provide an inappropriate framework for assessing the justice or injustice of the intersubjective meanings that comprise international institutions. As we shall see, the foundational norms of modern international society legitimize certain actors and forms of action and delegitimize others, thus affecting the background conditions in which international environmental accords are negotiated. If this leads to the marginalization of certain actors and their claims, then the resulting accords can hardly be considered just. But we cannot comprehend such injustice in distributive terms. Meaningful participation in the process of norm formation is neither finite nor quantifiable and cannot be divided or distributed, even if some of the consequences of the resulting norms can. Thus, to paraphrase Kratochwil and Ruggie, a disjuncture exists between the social ontology of international institutions and the distributive theories currently available to assess their ethical standing.

Societal Ontology and the Critical-Theoretic Conception of Justice

As we have seen, the need to define justice in distributive terms arises from an individualist social ontology. Once individuals are seen as atomistic, rational pursuers of their own conception of the good, and once society is understood as primarily a system of exchange relations, a distributive conception of justice logically follows. A conception of

[21] Friedrich Kratochwil, *Rules, Norms, and Decisions: On the Conditions of Practical and Legal Reasoning in International Relations and Domestic Affairs* (Cambridge: Cambridge University Press, 1989), pp. 22–23.

[22] Friedrich Kratochwil and John Gerard Ruggie, "International Organization: A State of the Art on an Art of the State," *International Organization* 40 (Autumn 1986): p. 765.

justice that overcomes the limitations and blind spots of distributive ethics must, therefore, begin with fundamentally different ontological assumptions. For want of a better term, I call these assumptions the societal ontology.

This alternative ontology consists of three general propositions that explicitly contrast with the underlying premises of the individualist ontology. First, individuals are inherently social in a strong sense of the word, that is, their identities, interests, and actions are constituted by the normative structure of the social world in which they live. In other words, culture assigns "reality to actors and actions, to means and ends," and it endows "actor and action, means and ends, with meaning and legitimacy."[23] Second, individuals are not autonomous sources of their own conception of the good. As Charles Taylor argues, "A social view of man is one which holds that an essential constitutive condition of seeking the human good is bound up with being in society."[24] Finally, society is best thought of as a complex of human practices and intersubjective meanings embedded in, and constitutive of, layered institutional norms, rules, and principles. Although exchange relations are clearly an important dimension of human interaction, they do not constitute the totality of social life.

This societal ontology points to a distinctive interpretation of justice, the critical-theoretical conception. If social institutions can endow certain actors and forms of action with meaning and legitimacy, and in doing so deprive other actors and actions of social sanction, then an appropriate conception of justice should focus on actors' access to the processes of institutional production, reproduction, and dissemination. In other words, when considering the justice of background institutions we should forgo the traditional emphasis on right distribution and concentrate instead on meaningful participation in institutional decision making. From such a perspective, valid social norms need to satisfy what Jürgen Habermas calls "the principle of universalization." In short, "*All* affected can accept the consequences and the side effects its *general* observance can be anticipated to have for the satisfaction of *everyone's* interests (and the consequences are preferred to those of known alternative possibilities for regulation)." Habermas takes pains to stress that the principle of universalization can only be satisfied in a situation of practical discourse between all concerned. This stipulation denies two alternatives. First, "a norm cannot be considered the expression of common interest simply because it seems acceptable to some of them [those concerned] under the condition that it be applied

23 Meyer, Boli, and Thomas, "Ontology and Rationalization," p. 21.
24 Taylor, "Distributive Justice," p. 292.

in a nondiscriminatory fashion." And second, a valid norm cannot be deduced from a fictitious "original position," such as that advocated by Rawls. Imagined choice situations, in which individuals are stripped of their interests and removed from real-time discourse, are the very antithesis of the process of moral argumentation underlying the principle of universalization.[25]

INTERNATIONAL SOCIETY AND THE JUSTICE OF ENVIRONMENTAL ACCORDS

As noted earlier, Shue usefully identifies the central questions of justice that need to be asked about the distributive aspects of international environmental accords. I have argued, however, that distributive theories, and the questions they generate, are not well suited to assessing the background conditions in which accords are negotiated. Because of their underlying ontological assumptions, distributive theories strain to comprehend the justice of the intersubjective meanings that constitute the institutional framework in which negotiations take place. I propose a critical-theoretical conception that seems better suited to issues of background justice. Starting from a societal ontology, this theory of justice focuses on decision-making processes and the opportunities for meaningful participation. From this standpoint, it is possible to identify three questions that need to be considered in any assessment of the institutions that frame international environmental negotiations:

1. Which institutions of international society have the greatest implications for the justice of international environmental accords?
2. How do these institutions affect actors' abilities to participate freely in the development of international environmental norms, rules, and principles?
3. What prospects are there for positive institutional change?

The Moral Purpose of the State: The Hierarchy of International Institutions

As indicated earlier, international society consists of four levels of institutions: issue-specific regimes; fundamental institutions, such as international law and diplomacy; the organizing principle of sovereignty, and the moral purpose of the state, its raison d'être. All of

---

[25] Jürgen Habermas, *Moral Consciousness and Communicative Action* (Cambridge: MIT Press, 1991), p. 65. For a critique and reformulation of Habermas's principle of universalization, see Agnes Heller, *Beyond Justice* (Oxford: Basil Blackwell, 1987), pp. 240–41.

these institutional levels have constitutive power: that is, they consist of norms, rules, and principles that "define the meaning and identity of the individual [or the state] and the patterns of appropriate economic, political, and cultural activity engaged in by those individuals [or states]."[26] Despite this common quality, however, international institutions can be ranked according to their relative constitutive importance. As Alexander Wendt and Raymond Duvall argue, "'higher' level (or more fundamental) institutions condition the establishment of 'lower' (or less fundamental) ones as they enable, make possible, or, in a different vocabulary, create the conditions of existence of, the latter."[27] There are two reasons why in our analysis of international environmental justice we should treat the moral purpose of the state as the most fundamental institution of contemporary international society.

The first reason is that the moral purpose of the state is in fact the primary constitutive institution of international society and cannot, therefore, be ignored. Among those interested in the constitutive power of institutions—the ability to give meaning and legitimacy to actors and actions—it is generally, and mistakenly, assumed that the principle of sovereignty is the foundational institution of international society. Wendt, for instance, claims that the point at which states embrace the principle of sovereignty—the paired ideas that the state is the highest authority within its borders and that it recognizes no higher authority outside those borders—marks the birth of international society. Before this moment states exist in a Hobbesian state of nature. "The principle of sovereignty transforms this situation by providing a social basis for the individuality and security of states."[28] But as I argue elsewhere, sovereignty is not an autonomous value, one without need of prior justification.[29] This is usefully illustrated through an analogy with individual independence. If I assert my right to independence, others, especially those affected by my declaration, are entitled to question the grounds on which I make this claim. This need for prior justification has not escaped modern philosophers interested in the individual's right to independence and freedom.[30] Independence is often portrayed as necessary for individuals to pursue their own life

---

[26] Meyer, Boli, and Thomas, "Ontology and Rationalization," p. 12.

[27] Alexander Wendt and Raymond Duvall, "Institutions and International Order," in *Global Changes and Theoretical Challenges: Approaches to World Politics for the 1990s*, ed. Ernst-Otto Czempiel and James N. Rosenau (Lexington: Lexington Books, 1989), p. 64.

[28] Alexander Wendt, "Anarchy Is What States Make of It," *International Organization* 46 (Spring 1992): 412.

[29] Reus-Smit, *Moral Purpose of the State*.

[30] For instance, John Rawls appeals to the prior good of self-respect to ground his argument for individual liberty. See Henry Shue, "Liberty and Self-Respect," *Ethics* 85 (April 1975): 197.

plans, which is in turn linked to notions of human dignity, self-worth, and moral agency. In other words, the value of individual independence has always had an underlying moral purpose.

The need to justify independence and autonomy in terms of another, more elementary value is greatly magnified in the case of the state, which is fundamentally reliant upon its ongoing legitimacy. States exist at the intersection of two normative systems: they are the principal members of international society, and they are the nodal points of domestic systems of rule. Existing at such an intersection, the legitimacy of the state is conditioned by two sometimes complementary, at other times conflicting, normative arenas. Legitimacy is always conditional, and both international society and domestic society grant states legitimacy on the grounds that they fulfill certain rules and meet given standards. Hence, state sovereignty cannot merely be asserted—it must be justified on some prior, socially recognized moral grounds. It is on these grounds that the legitimacy of the state and its continued sovereign rights depend. The value of state sovereignty must, therefore, always have an underlying moral purpose. Such purposes are embedded in the normative structure of international society, varying from one system of states to another. To ascertain the moral purpose of the state in a given international society we need to ask three questions: What is the prevailing ideal of the human agent? What is the relationship between human agents and society at large? And what role does the state perform in relation to these agents and societies?

The second reason why we should treat the moral purpose of the state as the most important institution of international society is that the modern state's raison d'être has been directly implicated in the environmental crisis we now face. The origins of the modern moral purpose of the state lie in the radical social, political, and economic changes of the late eighteenth century. The most important of these transformations were the rise of industrial capitalism and the advent of a new social ideology centered on the rights and interests of the individual.[31]

The convergence of these twin transformations generated a new ideal of the human agent, new ideas about their relationship to society and nature, and, most important, a new understanding of the role of the state. Each of these ideas is enshrined in Adam Smith's epochal

[31] See Louis Dumont, "The Modern Conception of the Individual," *Contributions to Indian Sociology*, no. 8 (October 1965), and *From Mandeville to Marx: The Genesis and Triumph of Economic Ideology* (Chicago: University of Chicago Press, 1977); Charles Taylor, *Sources of the Self: The Making of the Modern Identity* (Cambridge: Harvard University Press, 1989); and Stephen Toulmin, *Cosmopolis: The Hidden Agenda of Modernity* (Chicago: University of Chicago Press, 1990).

work *The Wealth of Nations*, published in 1776. Individuals, for Smith, are self-interested, acquisitive utility-maximizers, and the essence of human nature is "the propensity to truck, barter, and exchange one thing for another." Society is thus understood as a system of exchange relations. "Every man thus lives by exchanging, or becomes in some measure a merchant, and the society itself grows to be what is properly a commercial society."[32] As Louis Dumont observes, Smith's writings signal a crucial shift away from an image of society that values "relations between men" to an image celebrating "relations between men and things."[33] In such a context, the role of the state, its moral purpose, is defined largely in economic terms. In addition to defense, the state's two crucial functions, according to Smith, are the protection of property rights (which, like Locke and Hume, he defines as justice) and the maintenance of the conditions necessary for commercial society to flourish (such as the provision of public infrastructure, including roads, bridges, and canals, and public education for the lower classes).[34]

Thus from the late eighteenth century onward the moral purpose of the modern state has developed a distinctly economic profile, evolving over time from Smith's rather minimalist conception of what this entails to the expansive image of the developmentalist state. Even the nationalist ideal that the modern state exists primarily to defend the autonomy of a distinctive national community has only provided an enduring basis for state legitimacy so long as political elites have been able to deliver the trappings of modern, consumer society. Nationalists in Eastern Europe and the former Soviet republics are not indifferent to the type of independent states they seek to construct; they want to create modern, industrial democracies, taking the Western European countries as their models. Moreover, their ongoing political power will depend to a large extent on the degree to which they approximate this ideal.

The transformations of the late eighteenth century ushered in what Carolyn Merchant calls an "ecological revolution," a revolution in the relation between human society and nature.[35] The modern "economic" state has been directly implicated in this ongoing revolution in two ways. First, one of the original justifications for the modern state was its role in protecting the property rights of those who transform nature

---

[32] Adam Smith, *An Inquiry into the Nature and Causes of the Wealth of Nations* (Chicago: University of Chicago Press, 1976), pp. 17, 26.

[33] Dumont, *From Mandeville to Marx*, p. 4.

[34] Smith, *Wealth of Nations*, pp. 231, 244–309.

[35] Carolyn Merchant, *Ecological Revolutions: Nature, Gender, and Science in New England* (Chapel Hill: University of North Carolina Press, 1989). Also see her earlier work *The Death of Nature: Women, Ecology, and the Scientific Revolution* (San Francisco: Harper & Row, 1980).

by their industry. Since Locke, it has been generally accepted, at least until recently, that nature has little or no value until it is mixed with human labor and thus transformed into productive personal property, which it is the state's responsibility to protect.[36] This relationship between the state's role in protecting the rights to property that individuals derive from their exploitation of nature found explicit formulation in eighteenth- and nineteenth-century arguments justifying the state's role in European colonial expansion. The Americas and Africa in particular were portrayed as vast arenas of natural bounty lying fallow because the indigenous peoples lacked the industry and wherewithal to transform them into productive property. Europeans not only had the right to exploit this bounty, colonial states were obliged to protect the property rights of those explorers, pioneers, and merchants who released nature from its unproductive slumber.

Second, the state has itself become one of the most important agents of natural exploitation. Its role in providing commercial infrastructure has grown in the last two centuries from the construction of roads, bridges, and canals to involve some of the most ambitious transformations of nature ever undertaken. Many of these have been deliberate attempts to reconstitute nature, such as the extraordinary hydroelectric schemes of which China's Three Gorges project is only one example, but many others have mutated or destroyed the natural environment through their unintended consequences. The state's role in the development of highly centralized power generation grids, which inefficiently consume huge quantities of fossil fuels, has greatly contributed to the production of the greenhouse gases now implicated in global climate change. And nuclear power generation, it seems, has been viable only in countries where the state has sheltered the industry from the high costs of "excessive" safety regulation and public scrutiny.[37] The Chernobyl disaster highlights the impact that such initiatives and strategies can have for the natural environment and its human inhabitants.

The Moral Purpose of the State: Patterns of Inclusion and Exclusion in International Society

How does the moral purpose of the modern state affect actors' abilities to participate freely in the development of the norms, rules, and principles that make up international environmental institutions? As

[36] See Locke's arguments about nature, property, and the state in *Two Treatises of Government* (Cambridge: Cambridge University Press, 1988), pp. 286–302.

[37] For an excellent analysis of the state's role in the development of the nuclear industry, see Joseph Camilleri, *The State and Nuclear Power: Conflict and Control in the Western World* (Brighton: Wheatsheaf, 1984).

we have observed, institutions endow social, political, and economic actors and actions with meaning and legitimacy. Thus existing institutions influence participation in the construction of new institutions in two ways: first, they empower some actors and marginalize others, and second, they license certain moral claims and deprive others of meaning. The moral purpose of the modern state frames international environmental negotiations in both of these ways. States that assume the economic or developmentalist mantle constitute the legitimate actors in these negotiations, actors whose interests claim particular legitimacy. Furthermore, moral claims concerning justice or the environment have special salience when defined in economic or developmentalist terms. In contrast, actors who cannot or will not assume the garb of the modern state, or whose moral claims contradict or clash with the goals of economic rationalism, are effectively marginalized or excluded from the process of environmental institution building.

During the nineteenth century the moral purpose of the modern state came to define the boundaries between the expanding European state-system and the non-European world.[38] Gerrit Gong has shown how the European states employed a "standard of civilization" to exclude Asian, African, and American countries (except the United States and Canada) from membership in international society. Nineteenth-century international lawyers employed this standard to divide the world into civilized, barbarian, and savage countries, denying membership to all those who failed the test of civilization.[39] One of the most interesting aspects of this exclusionary technique is how civilization was defined. As Michael Adas explains, the Europeans understood civilization in terms of scientific and technological progress, which was in turn linked to economic dynamism. As industrialization spread, "European observers came to view science and especially technology as the most objective and unassailable measures of their own civilization's past achievement and present worth."[40] The barbarian states of China and India were thought of as once-great civilizations, sadly ossified under the weight of traditional social structures, and Africans were considered savages devoid of all scientific and technological inclination or ability. It was in terms of these standards that international society gradually expanded, and when it was finally

---

[38] The best work on this subject is *The Expansion of International Society*, ed. Hedley Bull and Adam Watson (Oxford: Clarendon Press, 1984).

[39] Gerrit Gong, *The Standard of "Civilization" in International Society* (Oxford: Clarendon Press, 1984).

[40] Michael Adas, *Machines as the Measure of Men: Science, Technology, Ideologies of Western Dominance* (Ithaca, N.Y.: Cornell University Press, 1989), p. 134.

globalized with the post-1945 wave of decolonization, the ideal of the dynamic economic, scientific, and technological state remained the standard of achievement, becoming embedded in the charters of key institutions such as the World Bank, the IMF, the OECD, and GATT.

A full analysis of how the moral purpose of the modern state affects patterns of inclusion and exclusion in the formation of international environmental norms is beyond the scope of this chapter, but it is useful to examine briefly the fortunes of three differently positioned groups of actors that offer suggestive lines for future research. The first group consists of networks of scientists, which Peter Haas and others have termed "epistemic communities."[41] If these scholars are correct, then epistemic communities have exerted considerable influence over the formation of international environmental norms. Because they possess, or are said to possess, consensual knowledge on a particular environmental problem, they are often able to raise environmental consciousness, define state interests, and facilitate international cooperation leading to positive environmental reform. Haas claims "that if epistemic communities exist, and if they maintain fairly stable access to decision-makers and keep rivals at bay, then international arrangements that closely resemble the community's preferences will develop and endure."[42] Given the extraordinary complexity of the natural processes involved in global environmental change, it would be absurd to deny the importance of scientific and technical knowledge and information for the international policymaking process. Nevertheless, as Sheila Jasanoff explains in Chapter 8 of this book, the ease with which environmental change has been transformed into a scientific issue is only partly explained by the need for policy-relevant information.

If we focus on how the moral purpose of the modern state legitimizes certain sorts of claims in the process of institution building, we can make two preliminary observations about the influence of epistemic communities. First, since the late eighteenth century the ideal of the progressive economic, scientific, and technological modern state

[41] Epistemic communities are defined by Haas and others as "transnational networks of knowledge based communities that are both politically empowered through their claims to exercise authoritative knowledge and motivated by shared causal and principled beliefs." For a good overview of this approach, see Peter M. Haas, "Obtaining International Environmental Protection through Epistemic Consensus," *Millennium: Journal of International Studies* 19 (Winter 1990): 347–65, and *Saving the Mediterranean: The Politics of International Environmental Cooperation* (New York: Columbia University Press, 1990). Also see *International Organization* 46, no. 1 (Winter 1992), a special issue titled "Knowledge, Power, and International Policy Coordination."

[42] Haas, "Epistemic Consensus," p. 352.

has been intimately connected with the parallel development of science as the dominant form of human knowledge.[43] It should come as no surprise, therefore, that as long as scientists can maintain a united front they will have uniquely privileged access to the policymaking processes of the modern state. Second, it is remarkable how the way in which the European colonial powers used assertions about their own scientific and technological superiority to define the boundaries of international society is echoed in current environmental negotiations. Developing countries have become increasingly concerned about how industrialized states have been using so-called objective scientific data to bolster certain global environmental policy prescriptions.[44] Overall, the underlying normative structure of international society, embodied in the moral purpose of the modern state, empowers the producers of scientific knowledge, namely, the epistemic communities, as well as the most privileged consumers of that knowledge, namely, the dominant industrialist nations.

At the other end of the spectrum we have the experience of indigenous peoples. In contrast to the tension between modern industrial, consumer society and the environment, indigenous peoples claim to have a more balanced, less-predatory relationship with the natural world, a relationship that many argue is crucial to the physical and cultural survival of both the tribal peoples and the survival of the ecosystems in which they live. In the words of an Amara-kaeri Indian: "Many people ask why we want so much land. They think we do not work all of it. But we work it differently from them, conserving it so that it will continue to produce for our children and grandchildren. Although some people want to take it from us, they then destroy and abandon it, moving elsewhere. But we can't do that; we were born in the woodlands. Without them we die."[45] Expressing similar sentiments, a statement by the Coordinating Body for the Indigenous People's Organizations of the Amazon Basin declares that "We, the indigenous peoples, have been an integral part of the Amazonian biosphere for millennia. We use and care for the resources of that biosphere with respect, because it is our home, and because we know that our survival and that of future generations depend on it. Our accumulated knowledge about the ecology of our forest home, our models for living

---

[43] See Stanley Aronowitz, *Science as Power: Discourse and Ideology in Modern Society* (Minneapolis: University of Minnesota Press, 1988).

[44] See Anil Agarwal and Sunita Narain, *Global Warming in an Unequal World: A Case of Environmental Colonialism* (New Delhi: Center for Science and Environment, 1991).

[45] Amara-kaeri Indian, "We Respect the Forest," in *The Indigenous Voice: Visions and Realities*, vol. 1, ed. Roger Moody (London: Zed Books, 1988), p. 211.

within the Amazon biosphere, our reverence and respect for the tropical forest and its other inhabitants both plant and animal, are the keys
to guaranteeing the future of the Amazon basin."[46] These claims suggest that many indigenous cultures hold views about the relationship
between humans and the natural environment fundamentally different
from those advanced by Western industrialized cultures.

Unlike scientific communities, indigenous peoples have been largely
unsuccessful in influencing the underlying norms of emerging international environmental accords. Industrialized and developing states
alike are constructing environmental agreements around the amor-
phous goal of sustainable development, a goal that has more in common with the values of economic and industrial growth and progress
enshrined in the moral purpose of the modern state than with indigenous values of balance and harmony between human activity and nature.[47] The ideals of the modern state so dominate the negotiating
agendas of the developing countries in which many indigenous peoples
reside—agendas reinforced by the priorities of international lending
agencies such as the World Bank and the IMF—that indigenous peoples have been forced to circumvent national decision-making processes by forming transnational groupings like the Coordinating Body
for Indigenous People's Organizations of the Amazon Basin. But in
seeking international support for their causes, indigenous peoples have
willingly or unwillingly had their cultural survival aligned with other
modernist values. To begin with, attracting outside interest and concern has frequently entailed the commodification of indigenous culture, its transformation into an artifact worthy of Western conservation efforts. Second, one of the most popular arguments for preserving
the native forests where many indigenous peoples live is that these
wilderness areas contain an abundance of natural "secrets" and resources awaiting exploitation by Western medical, biological, and genetic researchers. Hence, native forests warrant protection on utilitarian and scientific grounds. These values are in turn invoked to
justify protection of indigenous peoples, on the grounds that indigenous knowledge can help unlock the secrets of the jungle. In sum, the
hegemony of the moral purpose of the modern state effectively silences
the moral claims of indigenous peoples, and their only other options

[46] Coordinating Body for the Indigenous People's Organizations of the Amazon Basin,
"We Are Concerned," *Orion Nature Quarterly* 9 (Summer 1990): 36.
[47] See "Environment, Technology, and Ethics," in *Ethics of Environment and Development: Global Challenge, International Response*, ed. J. Ronald Engel and Joan Gibb
Engel (Tucson: University of Arizona Press, 1990), p. 27.

are to identify their survival with the utilitarian, scientific, and commodity values of modern society.

The final case that illustrates the inclusionary and exclusionary dynamics of the moral purpose of the modern state comes from unexpected quarters, from the center of European policymaking circles. In the first major study of the values underlying environmental policy formulation in Europe, Willett Kempton and Paul Craig conducted extensive interviews with environmental policymakers in Austria, Germany, Sweden, and the United Kingdom.[48] In a recent article discussing their findings, Kempton, Craig, and Glasser point out that environmental negotiations and debates are usually dominated by the use of objective language and appeals to utilitarian values. This compares with public discourse on the environment that exhibits "more heart felt, wider identifying, and ostensibly 'subjective' language that often reveals a belief in the intrinsic value of the environment." Their interviews revealed, however, that most of the European policymakers they consulted shared the subjective nonutilitarian values found in public environmental discourse. Unfortunately, the same policymakers admitted that these values could not be expressed in the official context of environmental negotiations. "Many interviewees stated that institutional pressures forced them to separate and sublimate their personal environmental views to satisfy institutional constraints."[49] The authors do not discuss the nature of these institutional pressures, but it is worth considering the extent to which the moral purpose of the modern state, with its associated economic, scientific, and technological imperatives, provides the basic system of institutional constraints and incentives. If this is the case, then the moral purpose of the modern state constitutes a horizontal mechanism of exclusion extending across international society, silencing counterhegemonic moral claims originating from both the core and periphery.

The Moral Purpose of the State: Prospects for Change

Most scholars who consider the prospects for meaningful international change in response to the challenges posed by the environmental crisis begin with the problem of state sovereignty. There are some who be-

[48] The conclusions of their study have so far appeared in Willett Kempton and Paul P. Craig, "European Perspectives on Global Climate Change," *Environment* 35 (April 1993): 16–20, 41–45, and Willett Kempton, Paul P. Craig, and Robert Glasser "Ethics and Values in Environmental Policy: The Said and the UNCED," *Environmental Values* (Summer 1993).

[49] Kempton, Craig, and Glasser, "Ethics and Values in Environmental Policy," pp. 2, 19.

lieve that sovereign rights represent the principal obstacle to the international reform needed to avoid global environmental collapse and to meet basic human needs. Moving beyond sovereignty is thus seen as a crucial prerequisite for genuine change.[50] There are others who consider state sovereignty to be an inescapable reality of the contemporary world, the unavoidable starting point of any strategy for change.[51] There are elements of truth in both positions. It is difficult to deny that the sovereign state's territorial rights have been exploited to justify unfettered abuse of the natural environment and to frustrate protection efforts, but it also seems reasonable to assume that in the foreseeable future states will continue to assert their rights to independence. Yet despite the value of these insights, their focus on state sovereignty misses the true locus of international change and thus the real prospects for transformation.

I suggest that sovereign rights per se are not the crux of the problem; rather, it is the economic moral purpose of the modern state underlying and sustaining these rights that has fueled the global environmental crisis and currently poses the greatest obstacle to the development of just and effective international environmental institutions. Moving beyond the moral purpose of the modern sovereign state is thus the key to genuine international change. But what are the prospects of such change? In addressing this question we need to consider the general conditions in which deep normative structures of international societies change, whether such conditions exist in our present circumstances, and where such change is likely to originate.

Despite my emphasis on the constitutive power of the foundational institutions of international society, it would be wrong to assume that the moral purpose of the modern state is immutable. Like all social structures, the meanings that constitute these deep-seated institutions are nothing more than the routinized practices of human agents. When these practices change, the normative structure of international society must also change. History suggests that such practices usually change—producing and sustaining new ideas about the moral purpose of the state—when a change in the social structure of accumulation is accompanied by the rise of a new social ideology. The moral purpose

[50] See Richard Falk, *This Endangered Planet: Prospects and Proposals for Human Survival* (New York: Vintage, 1971), and John Dryzek, *Rational Ecology* (Oxford: Basil Blackwell, 1987).
[51] As Hurrell and Kingsbury note in their introduction to *International Politics of the Environment*, "Environmental issues will still of necessity be managed within the constraints of a political system in which sovereign states play a major part" (p. 9). This is the same starting assumption adopted by Haas, Keohane, and Levy in *Institutions for the Earth*, p. 4.

of the Ancient Greek state was the product of what Max Weber calls a "consumer" economy combined with the values of civic humanism. The Renaissance Italians transplanted civic humanism into a fundamentally different social structure of accumulation, characterized by Weber as a "producer" economy.[52] This amalgam gave the Renaissance state its own distinctive moral purpose. As we have already observed, the eighteenth-century rise of industrial capitalism combined with the ideology of individualism produced the economic moral purpose of the modern state.

Whether or not these enabling conditions currently exist is an open question. Still, two parallel developments over the last twenty years warrant further reflection. The early 1970s witnessed the beginning of two ongoing global crises: one in the prevailing social structure of accumulation and the other in the dominant social ideology. Since the late nineteenth century, industrial capitalism has been organized around the principles and techniques of mass production, a mode of production commonly called "Fordism." The large-scale manufacturing processes that characterize this mode of production depend on a combination of cheap and secure energy reserves, constantly expanding markets, stable interest and exchange rates, and managed labor relations, each requiring distinctive forms of state support and regulation. After the Second World War industrialized nations, under the tutelage of the United States, sought to guarantee these conditions through the creation of the Bretton Woods system of international economic management. In the early 1970s the system experienced a major crisis. Energy prices skyrocketed, markets contracted, interest and exchange rates fluctuated widely, and labor relations deteriorated, with the net result being increasingly frequent periods of global recession. This crisis has encouraged many authors to conclude that we are witnessing the painful emergence of a new post-Fordist social structure of accumulation.[53]

Since the early 1970s we have also witnessed a challenge to the modern world's dominant social ideology. This challenge has been generated by the global environmental crisis itself. One of the crucial

---

[52] See Max Weber, *The Agrarian Sociology of Ancient Civilizations* (London: Humanities Press, 1976), p. 48.

[53] See Michael J. Piore and Charles F. Sabel, *The Second Industrial Divide: Possibilities for Prosperity* (New York: Basic Books, 1984); Robert Cox, *Production, Power, and World Order: Social Forces in the Making of History* (New York: Columbia University Press, 1987); Michel Aglietta, *The Theory of Capitalist Regulation*, trans. David Fernbach (London: Verso, 1983); and Alain Lipietz, *Mirages and Miracles* (London: Verso, 1987), and *Towards a New Economic Order: Postfordism, Ecology, and Democracy* (Oxford: Oxford University Press, 1992).

pillars of modern ideology is the idea that nature exists for human exploitation, that it has no value until it is turned to productive purposes by industrious human labor. As we have seen, it was this idea that informed early notions of possessive individualism and helped define justice in terms of property rights. But these ideas can no longer be asserted with the confidence mustered by Locke and Smith. The climatic dangers posed by global warming, the increased skin cancer rates produced by ozone depletion, the massive repeated flooding of the Ganges delta caused by deforestation in the Himalayan foothills, and the tragedy of Chernobyl all highlight the interdependence between humans and the natural environment. Many proclaim the inevitability of a technological fix, the prospect of continued economic growth, and the certainty of "business as usual," but many others call for a more balanced, less predatory relationship between humans and nature. These developments suggest that at the very least we are currently in a period of ideological reevaluation, the results of which remain unclear.

If these economic and ideological changes are to generate a new moral purpose for the state, then the impetus must come from below. As we have observed, the modern state is situated at the intersection of two normative arenas: international society and domestic society. A state's viability depends in large measure on the legitimacy it commands in both spheres. John Meyer convincingly argues that a powerful world polity exists within international society that legitimates particular state forms. The institutions that define the world polity confer "great and increasing powers on states to control and organize societies politically around the values (modern rationality and progress) established in world political culture."[54] In other words, the moral purpose of the modern state is reinforced by an exogenous world polity.

We must, therefore, turn to the domestic arena for possibilities of change. It is important here to remember Gramsci's insistence that state power only remains stable if the need for coercion is diminished by a legitimizing ideology that produces a social consensus capable of sustaining the existing distribution of power and authority. The economic moral purpose of the modern state operates in just this way. Gramsci argues that the only way to overturn a hegemonic ideology such as this is through a "war of position," whereby a counter-hegemonic ideology is constructed within civil society and eventually

[54] John Meyer, "The World Polity and the Authority of the Nation-State," in *Studies in the Modern World System*, ed. John Meyer and Michael Hannan (New York: Academic Press, 1980), p. 109.

undermines the dominant ideological formation. It is precisely this sort of process that produced the moral purpose of the modern state out of the ruins of Absolutism during the seventeenth and eighteenth centuries.[55] The likelihood of this occurring now is as yet unclear, but the evidence that European policymakers personally embrace non-utilitarian environmental values is reason for hope. It has been suggested that new social movements represent promising agents of change. For instance, Joseph Camilleri and Jim Falk argue that these "new social movements pose a significant challenge to the political economy of modernity."[56] If this is so, then it may indeed be possible to redefine the moral purpose of the modern state, especially given the economic and ideological transformations currently under way.

I BEGAN this chapter with a distinction between the background justice and the internal justice of international environmental accords, arguing that the individualist ontology underlying theories of distributive justice makes them inappropriate for assessing the background circumstances in which agreements are negotiated. These circumstances consist of institutional norms, rules, and principles that are not easily comprehended from a distributive standpoint. To overcome this disjuncture, I suggested that we adopt a critical-theoretical conception of justice based on a societal ontology that is more in tune with the nature of international institutions. This conception of justice led us to examine the moral purpose of the modern state as the foundational institution of international society, to consider how this raison d'être gave certain actors and moral claims privileged access to the process of institution building while excluding others, and to speculate about the prospects for meaningful international change.

This, of course, begs the question of how critical and distributive theories of justice should be combined to assess the ethical status of international environmental accords. The broad implication of my argument is that such assessments need to begin with the central question posed by the critical theory: namely, have prevailing normative institutions of international society enabled all affected actors' to participate effectively in the production and reproduction of emergent international environmental norms? Unless this question is addressed

[55] See Albert Hirschman's outstanding work *The Passions and the Interests: Political Arguments for Capitalism before Its Triumph* (Princeton: Princeton University Press, 1977).

[56] Joseph Camilleri and Jim Falk, *The End of Sovereignty? The Politics of a Shrinking and Fragmenting World* (London: Edward Elgar, 1992), p. 221. Also see Paul Wapner, "Politics beyond the State: Environmental Activism and World Civic Politics," *World Politics* 47 (April 1995): 311–40.

first, questions about the fair or unfair distribution of the benefits and burdens generated by such regimes are moot. If certain actors are excluded from the process of norm formation—or their claims are marginalized—then it is difficult to see how any distributive outcome can be considered just.

Given the nature of several emerging environmental regimes, there may also be a case for applying such standards to the internal justice of international environmental accords. Because of continuing uncertainties about the nature and extent of the global environmental crisis and the costs it is likely to generate, many have argued that negotiations should focus on establishing a basic system of rules and guidelines rather than set inflexible targets and make once-off allocative decisions. In many respects this is the logic behind the United Nations Framework Convention on Climate Change. The combination of basic rules and specific decisions will vary from one environmental agreement to another, but it is clear that assessing norms, rules, and principles will be an important factor in determining the internal justice of accords. We may have reason, therefore, to extend our application of the critical-theoretical conception of justice from the analysis of background institutions to terms of the agreements themselves. From this standpoint, we should focus on the rules of ongoing participation enshrined in new accords—who gets to participate and under what conditions? We should also consider the extent to which the moral purpose of the modern state is being reinforced and replicated by these rules. In other words, we should ask how the deep normative structure of international society is being sustained by environmental regime formation.

6

Impoverishment and the National State

JOSEPH CAMILLERI

OVER the last two centuries or more, successive waves of technical innovation, nurtured by the spread of capitalism and the associated ideology of modernity, have resulted in the periodic retooling and reorganization of production, the large-scale reshaping of transportation and communication systems, and profound changes to rural and urban life. The cumulative effect of these trends has been the transnationalization of trade, production, and finance or, to put it differently, an increasingly integrated global system of social and economic interaction. Two distinguishing characteristics of this still unfolding historical process are the multifaceted degradation of the earth's fragile ecological systems and the international stratification of power and wealth.

These two facets of the contemporary human condition, which may be more loosely referred to as *global environmental change* and *global impoverishment,* are now widely regarded by theorists and practitioners alike as likely to reach crisis proportions in the absence of timely remedial action. Several difficulties, however, stand in the way. Three are especially worthy of attention: first, in neither case is there an analytical consensus as to the roots of the problem; second, the two problems may be closely interconnected;[1] third, the modern state, often characterized as the world system's dominant political institution,

---

[1] The cumulative and mutually reinforcing impact of what I have described as "systemic" and "ecological" imbalances is discussed in Joseph A. Camilleri, *Civilization in Crisis: Human Prospects in a Changing World* (Cambridge: Cambridge University Press, 1976), pp. 22–29.

may be as much part of the problem as part of any prospective solution. This chapter is primarily concerned with this third difficulty.

Before proceeding to a detailed examination of state structures, functions, and capabilities, it may be useful to sketch in broad outline the dimensions of the twin problems of ecological degradation and socioeconomic inequality. Human modification of the environment in itself is not new. What is new is that such modification is now likely to have global and possibly irreversible consequences, which may in time threaten the long-term habitability of the planet. Considerable uncertainty, however, surrounds the extent and likely trajectory of the major environmental problems that the scientific community has recently identified. The threats posed by global climate warming, ozone layer depletion, acid precipitation, tropical deforestation, and loss of species diversity have become familiar enough, but the institutional implications of these threats remain ambiguous to say the least. Can these problems be resolved by existing institutions simply operating more efficiently, that is, using more sophisticated technological or managerial know-how? Or might it be the case that environmental change, precisely because it has assumed global proportions, is no longer amenable to traditional institutional practices, which have tended to evade negative consequences (e.g., through migration) or to transfer them either to other places (e.g., downwind, downstream, to other countries or other regions) or to another time (namely, to succeeding generations)?

While focusing our attention on the *global* and *long-term* consequences of human intervention, environmental change has also acquired a political immediacy in both time and space. The depletion and degradation of resources are already sufficiently advanced to have provoked intense conflict both within and across states. These conflicts are likely to become all the more acute in the absence of formal, legitimate mechanisms capable of mediating claims and apportioning costs and benefits.[2] Resources, which may have once been abundant or at least readily accessible, are becoming scarcer for a number of communities. Environmental degradation (in particular soil erosion) is contributing to the destabilizing mass exodus from villages to Third World cities and the equally explosive flow of refugees from one country to another.[3] The depletion of forests and soils, and as a conse-

[2] See Jim MacNeill, Pieter Winsemius, and Taizo Yakushiji, *Beyond Interdependence: The Meshing of the World's Economy and the Earth's Ecology* (New York: Oxford University Press, 1991), pp. 54–55.

[3] See Jodi L. Jacobson, *Environmental Refugees: A Yardstick of Habitability*, Worldwatch Paper no. 86 (Washington D.C.: Worldwatch Institute, 1988).

quence the diminishing capacity of the land to support rising populations, are playing an important part in the social, political, and military upheavals in such countries as El Salvador, Somalia, Ethiopia, and Sudan.[4] The lack of adequate water resources has already emerged as a politically sensitive issue in the Middle East, South Asia, and parts of Africa and Latin America.

These observations underscore the complex interconnections between ecological degradation and economic impoverishment. The extent of global poverty itself has been well documented and need not detain us long. According to one estimate hunger and poverty-related causes account for thirteen to eighteen million deaths each year. More than one-fifth of the world's population lacks the diet needed to sustain normal work, while half a billion people are too hungry to engage even in minimal activity.[5] It is calculated that the two billion poorest people in the world (some 40 percent of the world's population) account for less than 12 percent of global income. Even in Brazil, a country that has experienced rapid industrialization, some 70 percent of the population remains undernourished and as many as 50 million fall below the line of absolute poverty.[6] In all of this the inescapable fact is the uneven distribution of wealth and income both within and between national societies.

Perhaps the most striking mechanism of the distributive process, particularly in its contemporary global setting, is Third World external debt, which rose from $50 billion in 1970 to $1.3 trillion in 1990, by which date the net transfer of financial resources from the less developed to industrial countries was estimated at $20–30 billion a year.[7] There is more to Third World debt, however, than the global distribution of financial resources. Debt may also be seen as cause and effect of the ecologically problematic relationship between population and resources now confronting ever larger regions of the Third World.

Viewed in this light Third World debt is part of a much wider phenomenon that links the overuse of environmental resources with the daily struggle for survival experienced by a large fraction of humanity. The breakdown of ecological systems arising from overgrazing, overcropping, deforestation, soil erosion, and desertification cannot but compound the predicament of the poor who live in rural areas and

[4] See Jim MacNeill, "The Greening of International Relations," *International Journal* 45 (Winter 1989–90): 5.
[5] James Gustave Speth, "A Post-Rio Compact," *Foreign Policy*, no. 88 (Fall 1992): 149.
[6] See Roberto P. Guimarães, *The Ecopolitics of Development in the Third World: Politics and Environment in Brazil* (Boulder, Colo.: Lynne Rienner, 1991), p. 214.
[7] Speth, "Post-Rio Compact," p. 149.

depend on the land for their subsistence. The resultant poverty may not be new, but the absolute scale of the problem certainly is. More important perhaps, the phenomenon is for the first time truly global in the sense that it has both global origins (in the structure of the international political economy) and global consequences (namely, its impact on global ecology), which is not to say that local factors and outcomes are not an integral part of the equation.

Poverty may be a critical source of environmental degradation, but so is affluence. One need only think of the prevailing patterns of resource-intensive production and consumption to be found in affluent countries (as well as in the affluent sectors within poor countries). To give but one example, it is primarily changes in the world market (mediated by Western consumer preferences and the profit-maximizing strategies of national and transnational agribusiness firms) that account for the expansion of commercial production (that is, cash crops) and the progressive encroachment on peasant communities. Deprived of the best arable land, they are often compelled to migrate either to urban centers or to marginal, more fragile lands, the net effect of which is likely to be social and ecological impoverishment. The decline in per capita food production since 1980 in more than fifty less developed countries, in part connected to marked soil deterioration, is one clear indicator of this trend. Even the introduction of technical innovations in the Third World (e.g., Green Revolution rice varieties, insecticides, and pesticides), while substantially increasing production, may have unintended but nevertheless negative consequences for both people and the environment.[8] As for large and medium-size Third World cities, the transfer of polluting industries from the affluent North combines with the migration flow from the countryside to create severe and ultimately unsustainable stress on the physical and social fabric of urban life.[9]

Environmental degradation has undoubtedly assumed global proportions, yet its most palpable effects have been felt in the poorest regions of the world, for it is precisely here that falling commodity prices, a crushing debt burden, and reverse financial flows, combined with policies of structural adjustment and rising protectionism in many industrialized societies are driving the cultivation of marginal

[8] For a more detailed discussion of these interconnections, see Gilberto C. Gallopin, Pablo Gutman, and Hector Maletta, "Global Impoverishment, Sustainable Development, and the Environment: A Conceptual Approach," *International Social Science Journal* 121 (August 1989): 376–78.

[9] See Porus Olpadwala and William Goldsmith, "The Sustainability of Privilege: Reflections on the Environment, the Third World City, and Poverty," *World Development* 20, no. 4 (1992): 628–30.

lands and overgrazing of pastures. It is hardly surprising therefore that Henry Shue and others should draw attention to the implications of global environmental change for distributive justice. The ecological crisis and the socioeconomic breakdown experienced in many parts of the Third World can hardly be analyzed, much less remedied, in isolation from each other. As Alvaro Soto has put it, "They are the interconnected result of an inequitable world order, unsustainable systems of consumption and production in the North, and inappropriate development models in the South."[10] We may even go so far as to speak of a "socio-ecological crisis" in which the close interaction between the local and global scales and between the political, economic, cultural, and biophysical dimensions of human organization accentuate vulnerabilities and reduce the capacity for adaptable innovation through increased foreclosure of alternative solutions.

The Socio-Ecological Crisis and the World System

What, one may ask, is the function of the state in all of this? To what extent and in what way does the state contribute to, institutionalize, or even legitimize the socio-ecological crisis? From what has already been said it should be clear that no answer to the question is possible without first locating the state within a wider historical canvas. The state is after all both agent and victim of the ebb and flow of change, with the result that changing circumstances are likely to produce different kinds of state at different times and in different places. The unevenness and differentiation inherent in the process are a continuing source of tension and a catalyst for further change.[11] If the aim is to establish the relationship of the state to the socio-ecological crisis we have just outlined, then we must first situate the state within the overarching international order of which the socio-ecological crisis is but a symptom. To this end world systems theory may prove especially useful in that it offers us a conceptualization of world politics (and of the state) that attempts to make sense of the globalization of human affairs while remaining sensitive to its fragmenting as well as integrative tendencies. A range of recent contributors, from John Burton to James Rosenau and Immanuel Wallerstein, have sought in different ways to depict an increasingly integrated yet structurally differentiated world.

[10] Alvaro Soto, "The Global Environment: A Southern Perspective," *International Journal* 47 (Autumn 1992): 679.
[11] See Philip G. Cerny, *The Changing Architecture of Politics* (London: Sage, 1990), p. 104.

What gives Wallerstein's representation of the world system greater explanatory power is its historical perspective and detailed consideration of the structures and processes of economic internationalization. He identifies the relationship between capitalist production (for a competitive world market) and the system of territorial states as giving rise to a series of historical changes that plot the trajectory of the "world-economy."[12] The systemic model developed by Wallerstein and his associates is theoretically instructive in that it synthesizes three interacting relationships that are germane to our analysis: that between the state and the expansion of capital, that between the international division of labor (core-periphery-semiperiphery) and the world market, and that between states and varying configurations of power. There are, nevertheless, severe limitations to this analytical framework, not least its economic reductionism, which assumes that all forms of political community, notably the state, are ultimately governed by their relationship to the international division of labor. There is, moreover, inadequate integration either of cultural processes as they unfold within and between societies or of the politics of localities, movements, and communities whose identities, interests, capabilities, and modes of action are distinct from those of states.

The reality is that while state and market are inextricably intertwined and drawn further and further by the very process of transnationalization into a "structured field of action," the interconnections between the state system and the world economy are not entirely functional; indeed, they are often contradictory. In any case, the notion of the political must itself be widened to include not merely states and their agencies (which, in his otherwise incisive contribution, Chris Reus-Smit labels misleadingly "international society"), but also the range of subnational, supranational, regional, and international institutions and processes with a stake in the authoritative allocation of resources.[13] We may refer to this complex and pluralistic international architecture of power and authority as the *world polity*. Yet there is still more to the world system than the interaction between the world economy and the world polity (and their constituent parts). To enrich our understanding of the world system we need to adopt at least a three-dimensional perspective by intruding the notion of civil society,

[12] See Immanuel Wallerstein, *The Capitalist World-Economy* (Cambridge: Cambridge University Press, 1979), pp. 1–36, 152–64.

[13] The widening framework of international institutionalization has in recent years been subjected to considerable, if theoretically still inadequate, scrutiny. See, for example, Evan Luard, *The Globalization of Politics: The Changed Focus of Political Action in the Modern World* (London: Macmillan, 1990), pp. 163–91.

defined as the realm of autonomous activity, that is, public activity separate or at least distinct from that of the state. Under this category we may include extended families, clans, villages, local communities, cooperatives of various kinds, groups for leisure or charity, religious and ethnic organizations, indeed the whole gamut of voluntary associations formed to advance particular interests or objectives. These *communities,* which we may describe as the building blocks of civil society, can give rise to multiple, overlapping, and intersecting layers of political association.

So far as the contemporary period is concerned, the notion of civil society has particular resonance in much of the postcolonial world, in part because the state, despite the best efforts of nation-building, is often regarded as alien, as imposed by force (whether from within or without), as inimical to traditional networks of association. But the phenomenon cannot be understood purely as postcolonial resistance to modernity. Technological change, which is central to the socio-ecological crisis, by its very nature, fosters new connections and solidarities, while at the same time reviving older loyalties and identities.[14] Civil society, then, need not be conceived as coinciding exclusively or even primarily with national boundaries. Indeed, we may speak of an emerging international civil society that constitutes more than the sum of all national civil societies or even the interlinking of these societies.[15] The hypothesis of an international civil society does not imply that a single set of values or philosophical assumptions sustains the operation of the world system, but that the plurality of highly differentiated cultural interpretations have over time come to share one thing in common: they all, in different ways and to different degrees, relate to and seek to illuminate "the modern global circumstance."[16]

We may then depict the world system as that *fluid pattern of integration and fragmentation deriving from the deepening, at times contradictory interconnections between world polity, world economy, and international civil society.* In one sense the three levels of systemic in-

[14] For a fuller discussion of the rise of societal actors (sometimes referred to as social movements), see Joseph A. Camilleri and Jim Falk, *The End of Sovereignty? Politics in a Shrinking and Fragmenting World* (Aldershot, Hampshire: Edward Elgar, 1992), pp. 206–11.

[15] For a more detailed exposition of my notion of "global civil society," see J. A. Camilleri, "The State, Civil Society, and Economy," in *The State in Transition: Reimagining Political Space,* ed. J. A. Camilleri, Anthony P. Jarvis, and Albert J. Paolini (Boulder, Colo.: Lynne Rienner, 1995), pp. 216–19.

[16] This phase is indicative of the multidimensional processes of globalization and is central to the argument developed by Roland Robertson and Frank Lechner, "Modernization, Globalization, and the Problem of Culture in World Systems Theory," *Theory, Culture, and Society* 2, no. 3 (1985): 103–17.

teraction are closely linked and occupy the same transnational public place; yet each has a logic of its own and access to a different set of resources; none is able to shape the world system purely in its own image or according to its own logic. It is the coexistence and fluctuating relationship of these three levels which give the world system its distinctive profile at any given stage of evolution. Robert W. Cox suggests yet another component as integral to world order, namely, the "biosphere or global ecosystem."[17] Although environmental constraints are often treated as an aspect or outcome of the functioning of the world economy, a persuasive case can be made for treating the global ecosystem as conceptually distinct. It can be said to function in accordance with its own logic, which affirms the complex interconnectedness of ecological subsystems, and hence the integrity of global space and fragility of the global ecological balance.[18] Understood in this sense, the biophysical system sets its own limits on the functioning of civil society, polity, and market.

On the basis of this four-dimensional conception of the world system, the contemporary period may be characterized in terms of an increasingly integrated global economy (which coexists with intensifying competition, uneven development, and environmental degradation), coupled with the emergence of a "global culture" and "world public opinion" (which coexists with ethnic, national, religious, ideological, and other forms of segmentation). The net effect of these two interacting trends is to accelerate the interdependence of states and growth of international organization (that is, integration of the world polity), but also to redefine the function of the state and its relationship to the culture and institutions of global civil society on the one hand and the dynamics of the world market on the other.

The question arises: How might these general theoretical propositions be applied to an analysis of the twin problems of global impoverishment and global environmental change? As a partial answer to this question, we postulate uneven development and the externalization of environmental consequences to be distinguishing characteristics of the growth-oriented competitive dynamic, or *dominant logic*, that underpins the structure of the world economy and its relationship to the fragmented system of states. This dominant logic is closely related to Reus-Smit's "moral purpose of the state," except that the logic de-

[17] Robert W. Cox, "Multiculturalism and World Order," *Review of International Studies* 18 (1992): 161.
[18] For a useful analysis of the implications of ecological norms for international relations theory and practice, see Hugh C. Dyer, "Environmental Ethics and International Relations," *Paradigms* 8 (Summer 1994): 59–72.

scribed here is more complex than the Reus-Smit formulation suggests and, though internalized and articulated by the modern state, must be understood as central to the functioning of the world system.

In recent years this dominant logic has been strikingly evident in the structural adjustment policies supervised by the World Bank and the International Monetary Fund, which have required a large number of governments situated in the periphery and semiperiphery to balance their budgets by cutting public expenditure, eliminate trade barriers and social subsidies, tighten monetary policies, encourage exports, and remove restrictions on foreign investment. Peripheral and semi-peripheral states have implemented this program with varying degrees of enthusiasm, giving the privileged access to scarce land, water, credit, and technology at the expense of the needs of small farmers, especially with regard to soil conservation, reforestation, and food security. The negative social and ecological effects of these policies have been widely documented, not least in the African context.[19] But the general pattern characterizes many parts of the Third World. According to one assessment, "Timber exporting has denuded mountains, causing soil erosion and drying critical watersheds. Cash crop exports have depended on polluting pesticides and fertilizers. Large fishing boats have destroyed the coral reefs in which fish breed and live. Tailings from mines have polluted rivers and bays."[20] Complementing and reinforcing these environmental effects has been the continued skewing of land distribution and the resultant pressure exerted by marginalized peasants on already fragile ecosystems.

Where peripheral states have intermittently sought to resist the orthodoxy of structural adjustment, they have had to contend with the financial strictures applied by the major multilateral agencies or private lending institutions. They have also had to contend with the actual or threatened retaliation of metropolitan governments.[21] To give

[19] See Fantu Cheru, "Structural Adjustment, Primary Resource Trade, and Sustainable Development in Sub-Saharan Africa," *World Development* 20, no. 4 (1992): 497–512; also International Labour Organisation, *Structural Adjustment and Its Socio-Economic Effects in Rural Areas,* Report to the eleventh session of the International Labour Organisation Advisory Committee on Rural Development (Geneva: ILO, 1990); Bade Onimode, ed., *The IMF, the World Bank, and the African Debt Crisis* (London: Zed Press, 1990); Mort Rosenblum and Doug Williamson, *Squandering Eden: Africa at the Edge* (San Diego: Harcourt Brace Jovanovitch, 1987); and Hamid Tabatabai, "Agricultural Decline and Access to Food in Ghana," *International Labor Review* 127, no. 6 (1988): 703–34.

[20] Robin Broad, John Cavanagh, and Walden Bello, "Development: The Market Is Not Enough," *Foreign Policy,* no. 81 (Winter 1990–91): 150.

[21] For the purpose of this essay, I am using the terms "core" and "metropolitan" and the terms "peripheral" and "dependent" interchangeably.

but one example, when the Nigerian government attempted to protect domestic production of millet and cassava in the interests of greater self-sufficiency, the Bush administration threatened to use a GATT provision to overturn the ban.[22] The relative weakness or dependence of the peripheral state is also manifest in its dealings with transnational enterprises. Mining agreements, for example, have often included only vague or ambiguous references to environmental protection and stipulated few legal obligations on the part of the enterprise to maintain environmental standards or repair any damage that might occur.[23] These same external pressures coupled with the internalization of the dominant logic by Third World elites in Southeast Asia, Latin America, and elsewhere have resulted in the rights and habitats of indigenous peoples being subordinated to corporate development, whether in cattle ranching, logging, plantation agriculture, mining, oil exploration, or hydroelectric schemes.[24]

Its outward manifestations may have varied, often quite markedly, but the same dominant logic has governed the policies of newly industrializing or semiperipheral states, although, particularly in the East Asian experience, government regulations, incentives, and subsidies have circumscribed the free play of market mechanisms. In the case of Brazil, a "technobureaucratic-military alliance" has administered the exploitation of the Brazilian commons in accordance with a resource allocation strategy that has extracted a stiff environmental premium and marginalized a large proportion of the population.[25] South Korea and Taiwan have achieved more spectacular and consistent rates of economic growth, perhaps because of greater levels of state intervention, but with a substantial deficit in equity and ecological sustainability. China's market-oriented modernization program has increased agricultural and industrial output at the cost of high and rising income disparities in both urban and rural areas.

The pervasive influence of transnational capital over government policy is not, in any case, confined to peripheral states, but equally apparent in the individual and collective decision-making processes of core states. Whereas in purely formal terms the Montreal Protocol constituted an agreement among governments, in practice the interests of the relevant manufacturing and marketing enterprises were central

[22] Melvyn Westlake, "GATT Wars Threaten African Food Security," *African Recovery* 4 (October–December 1990): 14.
[23] United Nations Environment Program, *The State of the Environment*, prepared by Essam El-Hinnawi and Manzur H. Hashmi (London: Butterworths, 1987), pp. 168–69.
[24] James A. Swaney and Paulette J. Olson, "The Economics of Biodiversity: Lives and Lifestyles," *Journal of Economic Issues* 26 (March 1992): 16–17.
[25] Guimarães, "Ecopolitics of Development," p. 216.

to the negotiations about levels of permissible emissions, timelines, and allocation of responsibilities.[26]

There is, then, a good deal of evidence to support the view that the socio-ecological crisis and its diverse manifestations in the advanced, newly industrializing, and less developed regions is the casualty of the dominant logic that governs the interaction between the world economy and the state system. There is a danger, however, of overstating the dominance of this logic, erroneously characterizing the influence of the dominant institutions as totally pervasive. The developmental tendencies we have outlined, though mediated by core economic and political actors, are seldom free of tension or contradiction. To begin with, the core is not monolithic. The relative decline of U.S. hegemony and the corresponding rise of other centers of economic power (in particular Western Europe and Japan) mean that core interests are open to conflicting interpretation, depending on the geopolitical or geoeconomic vantage point from which they are perceived. Not surprisingly, core states have adopted diverging policies with respect to energy conservation, global warming, development assistance, and debt rescheduling, and this divergence often stands in the way of a common approach, as reflected in numerous international negotiations, most recently at the United Nations Conference on Environment and Development and in the complex negotiations surrounding the Uruguay Round.

Second, there remains, despite the continuing financial, technological, and commercial dependence of much of the Third World and the homogenizing influence of modernity, a marked disjuncture of interests and perceptions between elites in the core and elites in the periphery. Though the decline of Soviet power and the end of the cold war have deprived the periphery of the leverage it once exercised by virtue of the East-West conflict, the unrelenting competition that has emerged between the major capitalist centers, especially on issues of trade and technology, provides a new opening for states in the periphery and semiperiphery to advance their particular interests. Clearly, their bargaining position is likely to be enhanced to the extent that they can coordinate their positions through multilateral forums and institutions, be they regional or international in scope. The preliminary attempts to establish an international environmental regime have provided Third World governments with precisely such an opportunity.

At the 1989 session of the UN General Assembly, many in the Group of 77 sought a range of concessions on debt, technology trans-

---

[26] Soto, "Global Environment," pp. 695–96.

fer, trade, and other "economic development" issues in return for a consensus on the resolution to convene the United Nations Conference on Environment and Development (UNCED). Agreement on Third World demands was, it is true, postponed to a later stage, but the principle had already been established that international collaboration on such questions as greenhouse gas (GHG) emissions would depend on the satisfactory resolution of the problem of equity, by which was meant the transfer of financial and technological resources on a "preferential and non-commercial" basis over and above normal development funding.[27] The same principle was embodied in the amendments to the Montreal Protocol in June 1990, which established a fund to assist less developed countries to finance the incremental costs of introducing substitutes for CFCs. The fund, which was initially set at $160 million, would increase to $250 million once China and India adhered to the protocol. By this time several comparable funds were under consideration, including a World Atmosphere Fund, a Planet Protection Fund, and a global warming fund, as were a number of proposals for financing these funds, including national levies based on a fixed percentage of GNP and a range of pollution taxes.

But points of contradiction tending to weaken the authority and efficacy of the dominant logic were not confined to relations among states. They were equally apparent in relations between states and their respective civil societies. For many the "moral purpose" of the state could not be equated simply with economic growth measured in aggregate terms. Not surprisingly, in Brazil the authoritarian state had to contend with "the explosive growth of community-based organizations," including neighborhood groups, professional associations, ecological organizations, and thousands of Christian base communities.[28] Popular movements and citizen initiatives were also flourishing in the Philippines, while large and small campaigns focusing on such issues as agrarian reform, timber logging, industrial pollution, nuclear power plants, and dam construction were directly or indirectly challenging official development ideology in many parts of the Third World.[29] Numerous women's organizations, tree planting cooperatives, and village self-help experiments were striving to achieve food security and protect the natural resource base of their communities.[30] Though these initiatives emanating from civil society were far from capturing the

[27] See MacNeill, Winsemius, and Yakushiji, *Beyond Interdependence*, pp. 62, 77, 100–101; also Michael Grubb, "The Greenhouse Effect: Negotiating Targets," *International Affairs* 66 (January 1990): 78.

[28] Guimarães, *Ecopolitics of Development*, pp. 216–28.

[29] Broad, Cavanagh, and Bello, "Development," pp. 152–60.

[30] Cheru, "Structural Adjustment," pp. 501–2.

state, they were beginning to build significant links with educational, bureaucratic, and even military elites, as well as with international (governmental and nongovernmental) development agencies, which the state could not ignore.

The political assertiveness of civil society was, if anything, at a more advanced stage of development in industrial societies. Here the *societal production of technological risk,* coupled with relatively high levels of education and ready access to sophisticated means of communication, has given rise to a new set of sensibilities and over time to a range of social movements, exerting variable but substantial pressure over the policymaking process. The net effect has been to politicize technological change and subject technical options to greater public scrutiny.[31] The transnational networks of knowledge-based communities, often referred to as epistemic communities, are also contributing to the process,[32] although as Sheila Jasanoff rightly points out, their influence and the efficacy of their challenge to the dominant logic may have been overstated in the U.S. literature. The production of scientific and technical knowledge cannot be separated either from the political process or from the economic organization of society, and cannot therefore be expected to present a coherent alternative to the prevailing political order. Steven Yearley is right to insist that the drift toward scientific and technical professionalism by a number of environmental organizations may dilute the radical edge of the movement's analysis and program. Yet, valid though they may be, these qualifications do not alter the fact that the NGO community is placing additional limits on the state's freedom of action and, perhaps more important, on its ability to shape the contours of an emerging world public opinion.

We have thus far pointed to several tensions operating either at the core of the world system or in the relationship between core and periphery and between state and civil society. The state is after all expected to fulfill not one but several purposes. The project of modernity may place a premium on economic growth, but such a "purpose" does not obliterate individual and social needs for order, security, and identity. As a consequence, the state is exposed to contradictory pressures and the dominant tendencies of the world system are thereby weakened and deflected.

Complementing and reinforcing these tensions is the state's diminishing capacity to manage the complex interaction between the local and global levels of social organization. There is more to this than the

---

[31] The function of the "new social movements" and their response to technological risk is more fully discussed in Camilleri and Falk, *The End of Sovereignty?* pp. 211–21.

[32] See Peter M. Haas, "Obtaining Environmental Protection through Epistemic Consensus," *Millennium* 19, no. 3 (1990): 347–63.

limitations imposed by global environmental change on the political processes and institutions of national states. What is in question is the capacity of states, which are by definition boundary-maintaining systems, to monitor, interpret, and modulate the interaction between the social and biophysical effects of human actions, and compounded with this the reciprocal influences of local (national and subnational) and global (regional, international, and planetary) change. By way of illustration, we would cite the cumulative impact of such local actions as the elimination of indigenous cultures, immigration, production of toxic chemicals, and agricultural subsidies on global social, economic, and ecological processes; and conversely the local ramifications of the international relocation of industries, global technological changes (e.g., raw material and factors substitution), changes in global information and communication flows, and cultural homogenization by one or more dominant powers.[33]

It is precisely because of the more clearly perceived limits to the efficacy of state action that states are progressively transferring authority, functions, and resources to a diverse panoply of regional and global organizations. This semiconscious attempt to expand the world polity, whose most visible outcome in the slow but steady growth of the United Nations system has in recent years given rise to innumerable negotiations, conferences, and conventions (e.g., Convention on International Trade in Endangered Species, European Convention on Long-Range Transboundary Air Pollution, Convention for the Protection of the Ozone Layer, the 1989 Basel Convention).[34] To these the UNCED Conference added the conventions on climate change and biodiversity, the Rio Declaration (a statement of principles derivative of the 1972 Stockholm Conference), and an extensive action program, *Agenda 21*.[35] These agreements, it should be stressed, were not simply the by-products of the respective national bureaucracies of the signatories. In almost every case a range of UN agencies and other international organizations played an important part in the necessary collection, interpretation, and dissemination of data, in the negotiation of norms and rules, and in the establishment of monitoring and inspection regimes.[36] As the Rio Summit demonstrated, international institu-

[33] For a more comprehensive list of these reciprocal influences see Gallopin, Gutman, and Maletta, "Global Impoverishment," pp. 389–90.

[34] See Dixon Thompson, "Trade, Resources, and the International Environment," *International Trade* 47 (Autumn 1992): 754–55.

[35] Mark Imber, "Too Many Cooks? The Post-Rio Reform of the United Nations," *International Affairs* 69, no. 1 (1993): 55.

[36] There is a sense in which international institutionalization is both cause and effect of the internationalization of civil society. See Peter Bradbury, "UNCED and the Glob-

tionalization has paved the way for conferences, conventions, and agreements that have in turn, through a process of incremental creep, expanded the functions of existing institutions (e.g., UN Environmental Program) and established the framework for new ones (e.g., Sustainable Development Commission, Global Environmental Facility).

None of this says that international institutions have adequately offset the deficiencies of state intervention. To generalize, changes in the international political structure have not been able to clear the backlog of first generation environmental problems, let alone keep pace with the heavier social, political, and environmental costs associated with a more recent phase of economic "development." In a remarkably candid assessment, the 1982 Nairobi Declaration attributed the disappointing results of the Stockholm Conference to "inadequate foresight and understanding of the long-term benefits of environmental protection . . . inadequate coordination of approaches and efforts, and . . . unavailability and inequitable distribution of resources."[37] In many respects, these shortcomings were a reflection of the design of international organizations, which tended to embody two contradictory logics: the dominant logic underpinning the structure of the world system and the emerging logic encapsulated by the notion of equitable and sustainable development. Superimposed on this normative contradiction was the opposition of state interests, which often blocked effective institutional reform[38] and placed severe limitations on the mandates and budgets of existing UN and other agencies.[39] The net effect was a process of ad hoc international political transformation, grappling with symptoms rather than causes and lacking the normative consensus needed to devise complex formulas for the redistribution of costs and benefits.

REGIME CHANGE: SOME INSTITUTIONAL IMPLICATIONS

The interconnected problems posed by global environmental change and global impoverishment have greatly exceeded the explanatory po-

---

alization of Civil Society," in *United Nations Reform: Looking Ahead after Fifty Years,* ed. Eric Fawcett and Hanna Newcombe (Toronto: Science for Peace/Dundurn, 1995), pp. 207–17. For a more general discussion of the trend, see Camilleri and Falk, *The End of Sovereignty?* pp. 94–97, 143–46, 178–79, 186–87.

[37] See Mostafa Kamal Tolba, ed., *Evolving Environmental Perceptions: From Stockholm to Nairobi* (London: Butterworths, 1988), p. 9.

[38] MacNeill, Winsemius, and Yakushiji, *Beyond Interdependence,* p. 31.

[39] See Konrad von Moltke, "The United Nations Development Systems and Environmental Management," *World Development* 20, no. 4 (1992): 619–20.

tential of traditional state-centric theories of international relations. Partly in response to this intellectual void, though still wedded to the centrality of the state, regime theory has gained considerable currency, notably in the United States. Its initial purpose was to analyze the institutional framework governing international trade and monetary relations, particularly in the context of the United State's declining hegemony.[40] More recently, attempts have been made to apply the same theoretical approach to an examination of the interlocking sets of norms, rules, and procedures that govern international transactions concerning environmental issues.[41] Some have entertained the notion of even more specific regimes relating to such putatively self-contained areas as air pollution, ozone layer depletion, and global warming. The difficulty with many of these formulations is that they give insufficient attention to three closely related questions: the normative basis of regime formation, the dynamics of regime change, and the interconnectedness of regimes.[42]

If global environmental change and global impoverishment are closely interconnected outcomes of the dominant logic of the world system (with its emphasis on material wealth, competitive markets, productivist technical innovation, and globalizing and state-mediated competition), then regime change, at least as conventionally defined, may not be enough. Adequate responses to the socio-ecological crisis may require political space for a new logic, a new normative framework, and presumably a new or substantially modified set of institutions, rules, and procedures. Many have indeed referred to this potential transformation as implying a paradigmatic shift.[43] What must be stressed, however, is that paradigmatic shifts seldom occur overnight and are seldom complete or definitive. On the other hand, such shifts are precisely the watersheds that separate one historical epoch from another. There is much evidence to suggest that both social and political actors are beginning to grasp the contradictory pressures inherent

[40] Stephen D. Krasner, ed., *International Regimes* (Ithaca, N.Y.: Cornell University Press, 1983); Robert O. Keohane, *After Hegemony: Co-operation and Discord in the World Political Economy* (Princeton: Princeton University Press, 1984); Charles P. Kindleberger, *The International Economic Order: Essays on Financial Crisis and International Public Goals* (Cambridge: MIT Press, 1988).

[41] See, for example, Jessica Tuchman Mathews, ed., *Preserving the Global Environment: The Challenge of Shared Leadership* (New York: W. W. Norton, 1990).

[42] These remain no more than pointers for future research in a long and otherwise illuminating survey of the literature on international regimes: Marc A. Levy, Oran R. Young, and Michael Zürn, "The Study of International Regimes," *European Journal of International Relations* 1 (September 1995): 267–330.

[43] Dean E. Mann, "Environmental Learning in a Decentralized Political World," *Journal of International Affairs* 44 (Winter 1991): 304–6.

in intense economic and political competition and the mutually reinforcing prospect of social and environmental breakdown. Here, it is
worth noting that these two trends bear a striking resemblance to the
stages of "anomaly" and "state of crisis" that Kuhn characterizes as
constituting the prelude to a paradigmatic shift.

For a good many theorists, and perhaps not a few practitioners, the
rising probability of climate change offers the single most important
catalyst for regime change. It provides the added stimulus, in the wake
of recent progress toward international institutionalization on several
related fronts (e.g., Law of the Sea, Nuclear Non-Proliferation Treaty/
IAEA Safeguards, ozone regime) to establish the basis for effective co-
alition-building.[44] Oran Young suggests that a more demanding set of
conditions must be met before regime formation on this scale can have
much chance of success. He lists: (1) feasibility of contractarian interactions; (2) equity as a principal object of negotiation; (3) availability
of clear-cut solutions; (4) availability of effective compliance mechanisms; (5) exogenous shocks or crises; and (6) effective leadership capable of sustaining the momentum of negotiations.[45] At first sight,
these requirements may appear so taxing and comprehensive that they
preclude the possibility of successful international agreement.

Yet a less exhaustive list of conditions is itself likely to fall short of
the mark. Indeed, there may be a case for arguing that the list should
be made both more stringent and more comprehensive, especially if
regime change is to be understood as the transformation of the normative, institutional, and legal order needed to address not just one environmental problem, however global its scope or far-reaching its impact, but the complex web of problems posed by the socio-ecological
crisis as a whole. The rest of this chapter will be taken up with a
reformulation of the notion of regime change, with a view to highlighting the implications for institutional innovation and the evolving
structure and function of the national state. More specifically, attention will focus on five closely interrelated theoretical and practical requirements: (1) achievement of a new normative consensus; (2) integration of equity and ecology in economic decision making; (3)
redefinition of the social contract; (4) institutional conditions appropriate to just bargains; (5) a revised relationship between the state on

---

[44] For a detailed discussion of the conditions for regime change, in the context of
global warming, see Fen Osler Hampson, "Climate Change: Building International Coalitions of the Like-Minded," *International Journal* 45 (Winter 1989–90): 36–74.

[45] Oran Young, "The Politics of International Regime Formation: Managing International Resources and the Environment," *International Organization* 43 (Summer 1989):
365–75.

the one hand and the market, civil society, and the world polity on the other.

If regime change is to approximate a paradigmatic shift in the norms, values, and beliefs underpinning the structural logic of the world system, how might we conceptualize the new paradigm? Several studies have already attempted a taxonomy of the competing paradigms to be found in the marketplace of ideas. Timothy O'Riordan and R. Kerry Turner postulate four model paradigms: (1) the "Cornucopian" paradigm with its emphasis on economic growth and market competition; (2) the "technocentric" paradigm, which sees technical fixes in resource management as the key to *sustainable growth*; (3) "communalist ecocentrism," which posits limits to growth and to centralization of social and economic systems; and (4) "deep ecology ecocentrism," which places the claims and entitlements of the human species within the much larger context of nature itself.[46] Carolyn Merchant examines three approaches: (1) an "egocentric" ethic grounded in individualism and human dominion over nature; (2) a "homocentric" ethic based on notions of the collective good and the need to protect and conserve nature for human benefit; and (3) an "ecocentric" ethic, which focuses on the entire cosmos and assigns intrinsic value to nonhuman nature.[47] Clearly, these and other classifications have a good deal in common: they contrast the dominant paradigm (variously described as "cornucopian," "egocentric," "modern," or "liberal-materialist") with the three most compelling intellectual and political reactions to the experience of modernity: (1) the need to guide technical innovation and so ensure that its consequences are consonant with social goals; (2) the need for equity (equality) in the distribution of costs and benefits associated with economic development; and (3) the need to establish the integrity, stability, diversity, and balance of nature as the overarching goal of social organization.

It will be some time before these varied responses to modernity can in any sense give rise to a new normative consensus. There is reason to think, however, that neither technocentrism nor ecocentrism can of itself provide the philosophical basis, much less widespread societal acceptance, needed to achieve a paradigm shift. The technocentric view that "society" and "environment" can be subdivided into a series of concrete and distinct problems whose solution rests primarily on the contribution of science and technology flies in the face of cultural,

---

[46] Timothy O'Riordan and R. Kerry Turner, eds., *An Annotated Reader in Environmental Planning and Management* (Oxford: Pergamon Press, 1983), pp. 6–7.

[47] Carolyn Merchant, "Environmental Ethics and Political Conflict: A View from California," *Environmental Ethics* 12 (Spring 1990): 45.

political, and economic reality, which, to a large extent, shapes the trajectory of scientific and technological change.[48] Scientific inquiry *can* increase our understanding of the complex dynamics governing human interaction with the "total Earth system," and technical applications can be devised to prevent or mitigate undesirable global change. But as the analyses offered by Jasanoff and Yearley clearly suggest, neither science nor technology can set the ethical or institutional parameters within which policy choices must be made.

As for ecocentrism (also referred to as biocentrism), the need to preserve biotic integrity, important though it is, leaves too many questions unanswered. Wendy Donner rightly reminds us that affirming the value inherent in the biotic community is one thing, but establishing a hierarchy of value quite another. That consumption by the industrialized world and by urban elites in the Third World fuels the engine of environmental degradation is obvious enough, but the economic and political structures that give rise to prevailing patterns of production and consumption cannot be understood, much less remedied, by mere reference to the "laws" of ecology.[49] The biocentric ethic cannot provide the necessary and sufficient guidance for interpreting existing power relations or for articulating the outlines of a preferred institutional framework. If global ecological problems are closely connected to the disproportionate consumption of resources, then the cultural and political dimensions of regime, let alone paradigmatic, change become paramount.

All of this is not to suggest that the ethos of science and technique on the one hand or respect for nature on the other will not play a decisive part in the elaboration of a new paradigm. Rather, they will need to be integrated into a normative framework that is predicated on a pluralistic conception of human development, which, for our purposes, we may define as the process of change (sometimes loosely referred to as sustainability) in which the use of resources, the direction of investments, the pace and quality of scientific and technological advance, and institutional innovation are consonant with one another

[48] For an analysis of the shortcomings of the "mechanical paradigm," from which they derive the rationale for the "precautionary principle," see Peter M. Allen, "Evolution, Sustainability, and Industrial Metabolism," in *Industrial Metabolism: Restructuring for Sustainable Development*, ed. Robert U. Ayres and Udo E. Simonis (Tokyo: United Nations University Press, 1994), pp. 78–100. See also Samir Amin, "Can Environmental Problems Be Subject to Economic Calculations?" *World Development* 20, no. 4 (1992): 523.

[49] For a critique of the biocentric/anthropocentric distinction, see Ramachandra Guha, "Radical American Environmentalism and Wilderness Preservation: A Third World Critique," *Environmental Ethics* 11 (Spring 1989): 71–83.

and designed to enhance both the current and future potential for satisfying human needs, not least of which is the need to achieve and
maintain a balanced and *sustainable* relationship with other forms of
life. Here the needs, traditions, and perspectives of the non-Western
world must be considered central to any intervening dialogue.[50] Perhaps one of the more remarkable developments of the second half of
the twentieth century is the growing acceptance of this formulation,
despite innumerable (and inevitable) variations of emphasis, in otherwise widely disparate intellectual, cultural, and political settings.[51]

The national state is hardly the engine driving this embryonic normative consensus, which is not to say that it may not play a part in the
process. Different states, representing different political and cultural
traditions, different economic and geopolitical interests, and different
stages of economic development, may, at least indirectly, promote different elements or dimensions of the new normative framework. Theorists and policymakers alike are beginning to explore notions of comprehensive and environmental security, which in different ways call
into question state-centric and military notions of security.[52] At the
level of practice a few core states (e.g., in Europe) are beginning to
entertain the limited reorganization of the productive apparatus to ensure lower levels of pollution and more effective remedial environmental action. For other core states the principal preoccupation remains
retention of existing economic advantages, support for financial dereg-
ulation, and minimization of restrictions that might impede the free
play of corporate production and marketing strategies. For peripheral
states the emphasis is on the redistribution of the costs and benefits of
development (e.g., easing of the debt burden or more effective access
to new technology). These states may give the impression of a more
radical posture, particularly on issues of distributive justice, but even
when demands for a New International Economic Order were at their
height, the emphasis was much more on redistribution among than
within states.

The state's contribution, then, to a normative shift is at best limited

---

[50] For all its substantial limitations, biases, and inconsistencies the World Commission
on Environment and Development may be understood as an embryonic expression of
this new ethic. See *Our Common Future* (Oxford: Oxford University Press, 1987).

[51] See Rachel M. McCleary, "The International Community's Claim to Rights in Brazilian Amazonia," *Political Studies* 39 (1991): 691–707.

[52] See Erik K. Stern, "Bringing the Environment In: The Case for Comprehensive Security," *Cooperation and Conflict* 30 (September 1995): 211–38; also S. Dalby, "Security, Modernity, Ecology: The Dilemmas of the Post–Cold War Security Discourse,"
Alternatives 17 (1992): 95–134; and Joseph A. Camilleri, "Security: Old Dilemmas and
New Challenges in the Post–Cold War Environment," *GeoJournal* 34, no. 2 (1994):
135–45.

or partial, often grudging, and at times even negative. Yet states participate in and lend weight to this aspect of regime change, in part because they have to take account of the multifaceted pressures emanating from civil society (in its national and global dimensions), but also because, as we have seen, the limitations on their capacity to handle global functions compel them to extend the framework for international institutions. Over time, these institutions provide the arena for the slow but steady negotiation of a new normative framework, an arena which no state or group of states fully controls and in which global civil society is able to intrude with increasing frequency and efficacy.

Closely related to the redefinition of norms is the second requirement of regime change, namely, integration of justice and ecology in economic decision making and, most important, in the decision to invest, which, as we shall see, has vast implications for the structure and function of the state. Such integration, to be effective, must be evident at all levels of social and political organization, from the local to the global. More specifically, mechanisms must be instituted to perform two interrelated functions: (1) assessing the consequences of a proposed decision and feeding them back into the criteria that govern the decision before it is made; and (2) communicating the information on which these decisions are based from one level to another (namely, from the subnational, to the national, to the global). Without these feedback mechanisms, not only are governments, corporations, and international organizations likely to undervalue the social and environmental consequences of their decisions, but international agreements are likely to operate in a vacuum.

To illustrate, it is not possible for the international community to express an interest in the preservation of the Amazonian rain forest (because of the damage that the depletion of natural resources might inflict on the quality of life in the industrialized world) while at the same time implicitly or explicitly dismissing the problem of poverty in Brazil as a domestic responsibility. Neither of these two objectives is likely to make much headway unless they are integrated, both conceptually and institutionally, in the policies of the United Nations and its agencies, the IMF, the World Bank, foreign governments, and of course national and regional governments in Brazil itself. Agrarian reform, cancellation of debt, and conservation measures thus become different facets of the one integrated policy approach.[53] Such integra-

[53] See MacNeill, Winsemius and Yakushiji, *Beyond Interdependence*, pp. 79–80, 83.

tion, however, cannot by definition be achieved purely or even primarily by market mechanisms.

Similarly, in the case of GHG emissions, the varied pace at which they are to be reduced in different parts of the world, the infrastructure that must be established to facilitate reduced fossil fuel consumption in less developed countries, and the financial compensation that is offered to them to assist the development of alternative energy sources must be part of an integrated package that reverses the current net transfer of resources from the less developed to the industrialized world. Here again, there is much more to this than negotiation of international conventions and protocols. Such initiatives have to be complemented by corresponding action at the local, national, bilateral, and regional levels.[54] The Rio Summit represents, symbolically, the first concerted attempt at such global integration, although the negotiations that led to it and the agreements that derived from it, not least the language of *Agenda 21*, still reflect in large measure the structural fragmentation of the state system and the asymmetric distribution of financial resources in the world economy.

The complex interaction between global environmental change and socioeconomic impoverishment indicates, then, the need for synthesis in both intellectual and practical terms. A new interdisciplinary methodology is called for, one that depends on (1) the simultaneous consideration of ecological, social, economic, political, and cultural factors; (2) the multidimensional consideration of processes and phenomena across different temporal and spatial scales (short-term and long-term, local and global); and (3) a systemic approach that can integrate the dynamic, nonlinear interrelations between variables of different natures and scales that are relevant to the diverse set of processes under consideration.[55] Such a synthesis involves much more than the material aspects of these relationships or the measurement of greenhouse gas emissions, GNP growth, or productivity in this or that industry. It inevitably involves "'cross-issue bargaining' in both strategic and practical terms"[56] necessitated by the increasing interpenetration of national economies on the one hand and the progressive breakdown of local and regional ecosystems on the other.

This leads us to the third condition for regime change, namely, the

[54] See Gallopin, Gutman, and Maletta, "Global Impoverishment," p. 381.

[55] Mann, "Environmental Learning," p. 324.

[56] See Ian Burton and Peter Timmerman, "Human Dimensions of Global Change—A Review of Responsibilities and Opportunities," *International Social Science Journal* 121 (August 1989): 300.

need to reconceptualize the social contract and endow it with the necessary flexibility for spatial, temporal, and functional synthesis. One of the key questions here, as Henry Shue so incisively puts it, is whose preferences (or interests, or rights) are to be included in the economic or environmental calculus? For several writers a more inclusivist definition of political obligation is already discernible in the evolution of global society.[57] A new *global social contract*, made up of a series of discrete yet closely interlinked bargains, is slowly emerging, most obviously through the growth of international institutions and the numerous international protocols, conventions, agreements, and treaties in which states (or governments acting on behalf of states) are the contracting parties. The global contract, however, is also embodied in the cumulative knowledge generated by epistemic communities, in the vast and expanding literature on a global ethic, in an emerging world public opinion, and in the actions and demands of the diverse collective actors that comprise global civil society.

For the authoritative institutions spawned by the global social contract, the global ecosystem constitutes the physical space over which they exercise their jurisdiction. This is the hypothetical condition from which the present reality is far removed, yet significant steps have been taken in this direction. Important in this respect is the notion of the global commons, which has already been applied to four portions of the globe: Antarctica, the atmosphere, the deep oceans, and outer space. There remains, however, considerable contention about how the global commons is to be understood, and about the appropriate principles of ownership and stewardship.

There has, nevertheless, been a steady, at times painful progression toward the view that the benefits derived from the global commons must accrue to the entire human community and to future generations, despite the reluctance on the part of the more powerful economic and political actors to forgo the privileges inherent in their control of superior financial, technological, and military resources.[58] A case in point is the refusal of a few countries to adhere to the Law of the Sea, which upholds the principle that the underlying unity of the oceans requires effective global management regimes. The Law of the Sea Convention must, in any case, be seen as a halfway house. On the one hand, it has, with the establishment of the two-hundred-mile exclusive economic zones, placed an additional 35 percent of the oceans' surface under national control with regard to management of natural resources. On

---

[57] MacNeill, Winsemius, and Yakushiji, *Beyond Interdependence*, pp. 20–21.
[58] *Our Common Future*, p. 261.

the other hand, it has imposed on states an obligation to ensure that their activities do not injure the health and environment of neighboring states and the commons.[59]

The Law of the Sea, the Outer Space Treaty, and the Antarctic Treaty have in different ways and to varying degrees established the principle that a significant part of the planet is not subject to state appropriation by claim of sovereignty. What remains to be seen is whether that principle will be geographically extended over time to incorporate other ecologically or socially sensitive regions of the world, and whether the principle will in the meantime be translated into appropriate forms of institutionalization. The key question here is whether emerging international organizations, rules, and procedures will effectively enshrine notions of global responsibility and accountability, without at the same time allowing the ideology of global governance to be used as a weapon by core states (predominantly located in the West) to impose their preferences and priorities on the periphery. A new international regime based on the twin concepts of global social contract and global commons raises the possibility of contractarian bargains that avoid both the politics of domination (one-worldism) and the politics of paralysis and fragmentation (statism).

We must now look a little more closely at the institutional requirements of "just bargains," which we previously listed as the fourth condition of regime change. The simple proposition is that for bargains to be "just" they must address the problem of global change in ways that are congruent with the needs and circumstances of all contracting parties, not least those whose limited power and influence (those located at the periphery) would otherwise lead their needs and circumstances to be ignored. The issue here is the reconstitution of political space— as much at the local as at the national or global level—to allow for increased options for self-reliant adaptation. The most important aspects of institutional innovation, most of which have already been adumbrated, may be designated as follows: multilayered institutionalization, participation and accountability, access to information, and temporal interconnectedness.

Multilayered Institutionalization

International institutions, necessary though they are, cannot succeed in isolation. They are useful for setting standards, monitoring systemic trends, and overseeing the transfer of resources, but the task of implementing change and of developing the required production, technolog-

[59] Ibid., pp. 272–74.

ical, and administrative infrastructure must remain the primary responsibility of national, and to an even greater extent local and regional communities, for it is these communities which express and sustain the attachment to time and place. Decentralization, understood as the devolution of powers and functions to lower levels of government, is one important mechanism for opening the policy process to more effective public participation and accountability. It is doubtful whether William Ophuls's formulation of the small-scale, autonomous, steady-state society[60] will in the foreseeable future be within the reach of the world's major population centers, but as John Dryzek and others have suggested, a move toward more effective decentralization may be an important precondition "for more substantial institutional innovation directed toward ecological rationality."[61]

Decentralization may be said to create the political meeting point between justice and sustainability. If national governments and international agencies are to pursue just and sustainable development policies and practices, their mandates, budgets, and programs must have legitimacy where it counts most, that is among the local and regional communities they are meant to serve. What is envisaged here—a few signs are already pointing in this direction—is a process of multi-layered institutionalization, in which local, microregional, national, macroregional, supranational, and international institutions connect, intersect, and overlap, and multiple forms of citizenship (e.g., communal, national, global) sustain and are in turn sustained by different layers of political organization.[62] As Iain Wallace and David Knight clearly argue, attachments to place cannot be reduced to "the territorially defined state." But one can take the argument further, as does Will Kymlicka, and suggest that there is more to place attachment than the question of scale or territory. Human societies develop a profound sense of space-related identity through a common relationship with their physical environment, but above all through a shared history, which is sustained by and gives meaning to a common culture. If we accept that the boundaries of political communities are culturally defined, then it is not only nations (in the modern sense) or local and regional subdivisions that are entitled to self-government, but a range of other communities, not least indigenous peoples. For the institu-

[60] See William Ophuls, *Ecology and the Politics of Scarcity* (San Francisco: Witt Freeman, 1977), pp. 225–32.

[61] John Dryzek, *Rational Ecology: Environment and Political Economy* (Oxford: Basil Blackwell, 1987), pp. 246–47.

[62] See the declaration on *Global Environmental Democracy* published by the Centre for Science and Environment (New Delhi) in preparation for the UN Conference on Environment and Development and reproduced in *Alternatives* 17 (Spring 1992): 269.

tional infrastructure to have the desired levels of complexity and re-
dundancy, decentralization must be given both territorial and cultural
expression.

Participation and Accountability

One of the primary objectives of decentralization is to subject the design, implementation, and evaluation of "economic development" to
greater public scrutiny and participation. It is only through such a
mechanism that production decisions can be based on social needs
that have been publicly expressed and identified. But decentralization
is not enough, for the participatory ethic must inform not only local
projects and institutions but also the more complex calculations that
govern investment decisions made at the national and international
level. This is another way of saying that economic choices lie at the
heart of the political process and that, for example, the structural adjustment policies adopted by international financial institutions and
national governments or the greenhouse gas emissions set by international conferences are lacking in justice to the extent that they are not
subjected to a democratic process.[63] Where Third World elites are prepared to implement policies despite the severe social and environmental costs these inflict on the local population (e.g., Mexican economic
strategies during the 1980s),[64] they merely widen the gap between the
state and civil society and accentuate the tendency toward political
repression. To replace the free play of market forces with the diktat of
core governments or elites in peripheral states prepared to do their
bidding, or even international bureaucracies, would not greatly advance the cause of justice, for the cost/benefit calculus governing the
policymaking process would still not be made by those most directly
affected. That is why problems of environmental degradation, even
those that have reached global proportions, cannot be solved by simplistic demands for uniform environmental standards, much less the
unilateral enforcement of such standards, including the unilateral imposition of trade tariffs and other sanctions.[65] Precisely because policies in this area are bound to affect different societies and communities

[63] See Samir Amin, "Can Environmental Problems Be Subject to Economic Calculations?" *World Development* 20, no. 4 (1992): 528–29.

[64] See Daniel Goldrich and David V. Carruthers, "Sustainable Development in Mexico?. The International Politics of Crisis or Opportunity," *Latin American Perspectives*
19 (Winter 1992): 100–110.

[65] Such an approach is proposed by Duane Chapman, "Environmental Standards and
International Trade in Automobiles and Copper: The Case for a Social Tariff," *Natural
Resources Journal* 31 (Summer 1991): 449–61.

differently, the differential impact on the satisfaction of human needs (including the impact on nature) must be subjected to a negotiation that is inclusive of all affected parties.

Access to Information

If the decisions taken by international agencies, national governments, and private corporations are to take account of likely social and ecological consequences, then the efficacy of such decisions will in large measure depend on the quality of information on which they are based. The social and environmental impact assessment procedure, which has gained considerable if still insufficient ground in the decision-making processes of industrial countries, represents an early attempt to improve that quality of information. The large number of studies commissioned by governments as well as regional and international organizations to assess the extent and likely effects of climate change, loss of biodiversity, or desertification are further attempts to gauge the local and global impact of technological change. Many of these information-gathering exercises are, however, often deficient in that they do not adequately investigate the interplay between social and environmental impact, or between local, regional, and global effects. Other deficiencies arise from the limited dissemination of such information (access is often denied to those who are most directly affected), from the privileging of certain types of information (such as technical expertise versus local knowledge), and from the secrecy that surrounds the decision-making process itself (what decisions are made, who makes them, and with what criteria in mind). Where institutional capacities for collecting, interpreting, and disseminating information are deficient, the resulting asymmetries of knowledge are antithetical to the negotiation of just bargains.

Temporal Interconnectedness

Much of the foregoing has stressed the need to identify, institutionalize, and connect different political spaces, given that the justice of any decision will depend on incorporating the vastly different ground that different actors occupy. Much the same principle applies to different time scales. There is, on the one hand, the normative conception of a just and sustainable economic order and the distance, not least temporal, that separates it from existing institutional practices. This gap can be bridged only by a variety of organizational strategies reflecting different time scales that range from the short term to the long term. The issue here is not merely one of pragmatism (only small steps may be politically feasible at any given point). In certain circumstances, the relatively short-term needs of the poor, the landless, and the unemployed might have to take precedence over the claims of conservation, not as a matter of expediency but as a matter of principle. It follows that different strategies operating with different time scales may have distinct institutional and procedural requirements, which accentuates the need for some overarching structure or arena where the competing demands of these temporally differentiated strategies can be, if not fully reconciled, at least periodically negotiated.

Closely connected to this *strategic* principle is the incorporation of the future into the present. This has particular reference to micro- and even macroeconomic policy where the estimated costs and benefits of particular investments and projects often discount the future and where the cumulative impact of such practices tends to foreclose future options. It is worth noting, however, that this practice is not confined to environmental change. The foreclosing of options may have equally devastating implications for social and economic deprivation. The key here is to locate the political space where, at least in the longer term, separate and distinct institutions focusing on seemingly separate and distinct issues (e.g., deforestation, foreign debt, unemployment) can identify common origins and common ends and pursue interconnected strategies.

Much of what was implicit in the foregoing discussion of institutional functions and capabilities, at least so far as the "national" state is concerned, can now be made more explicit by considering the fifth dimension of regime change. Here we postulate four closely interrelated levels of analysis: the state as one institutional layer among others; the state as an instrument of institutional innovation; the state as agent and victim of economic restructuring; and the state as both initiator and subject of international regulation.

To argue that traditional notions of national sovereignty are being steadily overtaken by the realities of ecological and economic interdependence has become commonplace. Shared ecosystems and the global commons are already recognized as parts of the planet that fall outside national jurisdictions. Though UN resolutions and other international legal documents are still replete with references to the sovereign rights of states, there is increasing emphasis on the responsibilities or duties of states "to protect the global and regional environment" and "ensure that activities within their jurisdiction do not cause damage to the environment of other states." The Rio Declaration or "Earth Charter" includes among the "basic duties" of states: responsibility for "the survival/integrity/sovereignty of the earth," "equitable sharing of responsibilities and benefits," and "the protection of indigenous peoples." The sovereignty discourse, which has dominated the modern conceptualization of political space, is already undergoing a normative shift of paradigmatic proportions. This should not surprise us, for the national state is not a frozen geopolitical entity but an evolving phenomenon situated in time and place.

The web of dependencies and interdependencies mediated by technological and ecological change has sharpened the experience of vulnerability and given rise to diverse cultural and institutional responses. This is not to imply that state-centered nationalism is declining, that the physical force available to the state has diminished, or that new universal structures are around the corner. The state continues to perform important economic, administrative, and diplomatic functions, but these must not be confused with sovereignty. The image of a world where space is appropriated and exclusively controlled by sovereign national states is giving way to new images that allow for a variety of spatial forms, not only for national, but for local, regional, global, and even functional space.[66] Increasingly, the national state has to operate within a complex mosaic of spaces characterized by multiple allegiances, new forms of identity, and overlapping tiers of jurisdiction.

It is not simply, however, that the state no longer exercises exclusive control over political space, but that it is directly contributing to the process of institutional innovation. Reference has already been made to the many international conferences, conventions, and protocols that have resulted from the intermittent, often ad hoc responses of states to the plurality of challenges posed by the socio-ecological crisis. The slow but steady expansion of UN agencies, functions, and budgetary resources, which the UNCED process will undoubtedly accelerate, attests to this same trend. The key test of their efficacy or adaptiveness will be whether they can sufficiently modify the functioning of such pivotal international organizations as the World Trade Organization, the World Bank, and the IMF and help to create a new framework that can facilitate contractarian bargains on a global scale.

The broadening and strengthening of institutional capacity is not limited to the international arena. Regionalization is another trend that has gained considerable momentum over the last two decades, with varying degrees of support from core and peripheral states. Apart

---

[66] For a more detailed discussion of this trend, see Camilleri and Falk, *The End of Sovereignty?* pp. 236–57.

from the legitimacy extended to the process by frequent UN declarations and resolutions, states have been attracted to regional institutionalization by the prospect of more immediate results and greater commonality of interests. Not only are regional or select-group agreements easier to achieve than agreements at the global level, but they can often take advantage of the resources and infrastructure of existing regional groups (e.g., European Union, ASEAN, APEC, South Pacific forum, Caribbean Community), which can in turn facilitate the achievement of global agreements.[67] Regional agencies to control air, river, or marine pollution from the waste products of energy production are beginning to emerge, as are proposals for regional cooperation among states sharing common borders or similar socioeconomic and bio-climatic conditions.[68] Such agencies can foster joint action on energy, environment, and development problems; compile regionally comparable data on such issues; and institute surveillance, monitoring, and early warning capabilities that can in turn prevent or reduce an increasing range of both local and regional environmental and developmental hazards.

The other side to this coin is localization or microregionalization, that is, the development or restoration of local institutions that are informed by local conditions, sensitive to local needs, and involve local communities in the planning process. Development for basic needs, mediated by local institutions, encourages local resource management, common property arrangements, and greater self-reliance in food production, often with far-reaching implications for existing forms of land tenure. To this extent, Third World governments closely linked to local landlords or foreign agribusiness interests may be reluctant to foster the process of localization and may even forcefully oppose it. Increasing international economic integration may have a similar impact in the industrialized world, although the pressures emanating from civil society for greater local autonomy are likely to gather pace.

In all of this the key to future development will, for some time to come, remain the relationship of the state to the economy. The question here is whether or not the state, either alone or in concert with other institutions, can establish a significant degree of societal control over economic and technological change. Sometimes environmentalists

[67] This possibility is canvased with specific reference to target-setting agreements on greenhouse gas emissions by Michael Grubb, "The Greenhouse Effect: Negotiating Targets," *International Affairs* 66, no. 1 (1990): 87–88.

[68] See *Energy 2000: A Global Strategy for Sustainable Development,* A Report of the World Commission on Environment and Development (London: Zed Books, 1987), pp. 58–59.

will frame the question in terms of "correcting perverse interventions in the market,"[69] whether it be energy subsidies that favor large supply projects and undermine funding for energy conservation and renewable energy technologies or tax concessions for logging or mining that accelerate deforestation, species loss, or soil and water degradation.

These and other proposals, including the greening of subsidies, the reform of tax and tariff structures in the transport sector to encourage rail transport, and environmental taxes to promote end-use efficiency and mitigate the dangers of global warming,[70] though often couched in the language of free market principles, are predicated on a much greater level of state intervention and regulation. The setting, monitoring, and enforcement of standards, the imposition of green taxes and subsidies, or the introduction of waste management programs are all forms of intervention that create some markets, while destroying or modifying others. The same is true of the numerous recommendations periodically made (some radical, others less so) for international taxes to reduce disparities of wealth and income. If there are to be significant changes in the volume and quality of development assistance, the pattern of trade and investment flows, or the structure of financial transfers (including debt rescheduling and "forgiveness"),[71] then presumably both national markets and the global marketplace will have to be substantially reorganized and regulated. Do national states, even those situated at the core, have the infrastructural capacity, let alone the will or the legitimacy, to undertake such a comprehensive program of ecosocial engineering?

The answer to this question is neither simple nor definitive. Part of the answer must be that the national state still commands sufficient moral and legal authority and sufficient material resources to exert a profound influence on economic decision making. But the transnationalization of trade, production, and finance is now at such an advanced stage that the task is beyond the isolated efforts of metropolitan, let alone peripheral, or semiperipheral states. By way of illustration, the Mexican Ministry of Urban Development and Ecology had nine inspectors available to it to enforce environmental laws applicable to 35,000 metropolitan industries.[72]

To give one other example, it has been argued that to minimize the

---

[69] MacNeill, Winsemius, and Yakushiji, *Beyond Interdependence*, p. 33.

[70] Ibid., pp. 37–39.

[71] See for example, Speth, "Post-Rio Compact," pp. 155–59; also Ernst U. von Weizsäcker, "Sustainability: A Task for the North," *Journal of International Affairs* 44 (Winter 1991): 422–23.

[72] Goldrich and Carruthers, "Sustainable Development in Mexico?" pp. 105–6.

potentially dangerous social and ecological effects of biotechnology, the following measures would be needed: a capacity to monitor and control new technologies *before* they are tested and introduced; strict regulation of the interstate transfer of genetically modified organisms; regulation to prevent excessive uniformity of plant and animal varieties; a universal ban on the use of biotechnology for military purposes; adequate support and compensation for farming communities adversely affected by the introduction of new technologies; an early warning network to monitor the socioeconomic impact of biotechnology and its effects on biodiversity; and an international code of conduct to regulate biotechnology at all levels.[73] What state or group of states would be capable of performing this complex set of functions? The key issue here is not one of brute physical force, which states have in plentiful supply, but of organizational competence, spatial and temporal flexibility, intellectual coherence, and cultural legitimacy. In this sense the national state may lack the regulatory capacity needed to arrest, let alone reverse, global environmental degradation and global impoverishment. The actions of states must be complemented by the coordinated participation of a great many other actors, representing diverse interests and occupying diverse political spaces. State cooperation resulting in the further development of international law and international organization is necessary but not sufficient.

If economic and technological change is to be governed by criteria derived from a new normative consensus, then the state cannot be expected to be the prime mover in this process. A more likely outcome is that the state will function as the principal arena of conflict, a highly visible stage on which a range of competing values, interests, and organizational principles will contest the right to shape the emerging global political and economic order. How that contest unfolds, what particular form such globalism takes, will in large measure depend on the complex interaction between the world polity (which includes but is not reducible to the system of states) and international civil society. It is unlikely that a new global civilization equipped with a comprehensive system of authoritative institutions capable of meeting the challenge posed by the socio-ecological crisis will emerge in the near future. On the other hand, it is entirely possible that the praxis of social and political movements, coupled with the twin processes of political integration and fragmentation, will set the stage for a cultural and institutional pluralism more attentive to the implications of global impoverishment and environmental breakdown.

[73] See *Global Biodiversity Strategy*, prepared by the World Resource Institute, the World Conservation Union, and the United Nations Environment Program, 1992, p. 47.

7

Social Movements, Ecology, and Justice

SMITU KOTHARI

ACROSS much of the planet, the developmentalist regime of the past five decades has sharpened the contradiction between those who have benefited from it and those who have become its victims. Structural and social inequity, both within and between nations, and the politics of natural resource use are at the core of this polarization. Stated differently, in a stratified world, natural resource extraction continues to be controlled by and primarily benefits the privileged few, resulting in the increasing marginalization of millions dependent on these natural resources not only for their subsistence but also as a source of their identity. Global cultural and economic hegemony is flattening diversity—biological and cultural.

Throughout the Third World and also in industrialized countries, unequal control over productive assets continues. In India in 1990, for instance, the bottom 20 percent of people owned less than 1 percent of the assets, while the top 10 percent owned more than 50 percent. This reality is mirrored across the world in both the industrialized and less industrialized countries. Another illustrative case is the 60 million tribal people in India, who come from 212 different cultural groups. They are among the worst victims of uneven development. Most of the mining, hydroelectric, and industrial development has been implemented on their lands.[1] Planned development has, since Independence, forcibly displaced 15 percent of the entire tribal populace, making

[1] Ghanshyam, "Sustainable Development: Going Back to the Roots" (Madhupur, Bihar: Lok Jagriti Kendra, 1992, mimeographed); Janardhan Rao and G. Hargopal, "Adivasis in India: Transition and Development," *Lokayan Bulletin* 8, no. 6 (1990): 39–57.

them roughly half of the 20 million people who have been displaced since 1947.[2]

This scale of forced displacement raises a host of moral and ethical questions, particularly since states use the common "trade-off" justification or invoke "public purpose" or "national interest" in their defense.[3] In addition to the issues that this raises for the discourse on justice (see Will Kymlicka's and Henry Shue's chapters in this volume), other questions gain importance: Who decides what comprises the nation? Who benefits from "development" and who loses? Who decides *what* development is? Do displaced communities have no moral or justice claims? As India's former commissioner of scheduled castes and tribes, B. D. Sharma, has so cogently asked: Is the social identity of a people negotiable?[4] On what basis can we continue to justify the displacement, disorganization, and destitution of communities? In short, for how long can cultures and ecology—the sources of subsistence and identity—be sacrificed to sustain the affluent lifestyles and greater comforts for some and to create unsustainable material aspirations for a majority?

It is neither war nor social conflict, famines nor floods, that are causing this massive dislocation, but the dominant patterns of economic development themselves. Market rationality and profit maximization "discount the future" and are fundamentally incompatible with ecological sustainability and social justice.[5] Ecologically, the current strategies for achieving economic growth significantly depend on the extraction and utilization of natural resources far in excess of their capacities of natural regeneration. The market economy, with its continuing emphasis on maximizing productivity and profits, displaces millions from their economic and cultural moorings. Even if this could be justified, the dominant mode of industrialism is increasingly incapable of providing alternative livelihoods to more than a fraction of those displaced.

In addition, there has been a steady erosion of the control that individuals, communities, and nations can exert on what they produce,

---

[2] W. Fernandes and E. G. Thukral, eds., *Development, Displacement, and Rehabilitation* (New Delhi: Indian Social Institute, 1989).

[3] Smitu Kothari, "Whose Nation Is It? The Displaced as Victims of Development," *Economic and Political Weekly* (June 1996).

[4] B. D. Sharma, "Scheduled Castes and Tribes: A Status Report," *Lokayan Bulletin* 8, no. 6 (1990): 29.

[5] Smitu Kothari, "Challenging Economic Hegemony: In Search of Sustainability and Justice," Working Paper Series on *Development at a Crossroads: Uncertain Paths to Sustainability*, no. 11, Global Studies Research Program, University of Wisconsin, Madison, 1994.

how they produce, how what they produce is distributed, and who benefits from this production. Rather than decentralizing and democratizing control, there has been an increasing centralization of economic power within a small group of institutional and individual actors. The emerging global nexus between multilateral banks, transnational corporations, and global and national economic and political elites working through legal instruments like the World Trade Organization (WTO) threaten to recolonize much of the planet.[6] In fact, regimes like WTO, though seemingly multilateral, are bound to further capitalist accumulation by these elites as well as transnational corporations, thus compounding the already severe inequities both within and between nations. These processes and institutions remain overwhelmingly unaccountable and undemocratic.

A growing number of citizen's groups have, after a decade of effort around the world, demonstrated that while these institutions attempt to incorporate more socially and environmentally sensitive programs, they are structurally unwilling (or unable) to transform their basic economic agenda. It is indeed true, as Joseph Camilleri argues elsewhere in this book, that the core economic and political actors are not monolithic and that elites are marked not only by differentiation but also by "disjunction." India and Malaysia, for instance, played an important bargaining role in the preparations for the United Nations Conference on Environment and Development and in the final round of the negotiations of the World Trade Organization. Also, within each country, there are divergent interest groups that have not acquiesced to the dominant logic of industrial capitalism and the state system. Yet these divergences and disjunctions do not present an adequate challenge to the predatory interests of transnational capital or the hegemonic interests of the dominant capitalist centers. In fact, most industrializing countries have been compelled or deluded by the dominant economic institutions and their own elites (many of whom have been courted by hegemonic actors with the explicit intent of securing postcolonial access to labor, markets, and resources) into believing that they can "catch up with the West"—a prospect that, given finite natural resources and global structural inequities, is impossible to achieve. This iniquitous access and control over the Earth's resources along with the

[6] Smitu Kothari, "Global Economic Institutions and Democracy—A View from India," in *Beyond Bretton Woods: Alternatives to the Global Economic Order*, ed. John Cavanagh (New York: Pluto Press, 1994), and "Challenging Economic Hegemony"; Philip McMichael, "World Food System Restructuring under a GATT Regime," *Political Geography* 12, no. 3 (1994): 198–214; Chakravarti Raghavan, *Recolonization* (London: Zed Books, 1991).

large-scale dispossession and displacement of communities from their traditional sources of subsistence and meaning has engendered growing popular mobilization and social conflict. There can be little peace or security or democracy in the absence of a fundamental restructuring of these institutions and the developmental processes that perpetuate ecological degradation and social injustice. Such a transformation is essential if lasting social and economic justice is to be achieved. An ecologically sensitive analysis demonstrates that there is a basic incompatibility between intensive economic development of natural resources and the wider realization of distributive justice—a fact brought out by most mass-based social movements for economic and social justice.

These social movements, therefore, represent collective challenges to the dominant notions of "doing" development. They demonstrate (both through their resistance to current practices and by defining more sustainable and just alternatives) how development is a product of a particular historical configuration of power relations. Many have argued that "underdevelopment" is not a natural fact but an imaginary geography created by the "developed" world.[7] By essentializing as distinct categories "developed" and "underdeveloped," developed countries have defined the underdeveloped in a way that permits them to continue to control and manage the latter's affairs. Many critical voices in the social movements in the Third World challenge this hegemonic categorization further, arguing that this hierarchical perspective devalues much in their societies that is far more "developed" than the values and institutions of the "West." It is these voices from the margins of the world that are not only redefining democracy but are also raising some vital new questions. Below, we reflect on these questions and what they mean for the discourse on ecology, democracy, and power.

The Unfolding of Movements and Their Discourses

From Brazil to India and from the Philippines to Venezuela, ecology, survival, identity, autonomy, and justice are the main issues being raised by grassroots movements today. These struggles are challenging

[7] Jonathan Crush, ed., *Power of Development* (New York: Routledge, 1995); Arturo Escobar, *Encountering Development: The Making and Unmaking of the Third World* (Princeton: Princeton University Press, 1995); Gustavo Esteva, "Development," in *The Development Dictionary*, ed. Wolfgang Sachs (London: Zed Books, 1992); Jagannath Pathy, *Anthropology of Development: Demystifications and Relevance* (New Delhi: Inter-India, 1986).

two key models of our times: development and revolution. Unlike traditional working-class movements and/or peasant movements, particularly those inspired by socialist politics, these movements have no commonly accepted ideology, though the ground for convergence among them is steadily growing. There is a visible shift from emphasis on taking over the state to changing the processes of micropolitics itself.[8] In the context of Latin America and the Caribbean, Charles A. Reilly states that "social movements are having a profound structural and policy impact on (their) political systems."[9] Led by a wide range of new actors—indigenous peoples, forestdwellers, displaced persons, slumdwellers, workers in the "informal" sector, rural women—these movements are emphasizing a political and environmental agenda significantly different from that followed by those who were at the forefront of the struggles for justice in the past century. They are contesting the very thrust of contemporary economic development. These powerful economic assumptions, adopted by regimes as diverse as the United States, Singapore, and China, are being challenged on the grounds that they devalue nature, are unsustainable, and legitimate social injustice both within and outside the national boundaries of these countries as they scour other territories in search of natural resources and more pliant and cheaper labor. Movements argue that the problem of ecological imbalances is not primarily because resources are finite, but because economic development unsustainably extracts and pollutes these resources and rarely places limits on human consumption.

It must be noted that not all movements are intrinsically ecological. Not only does a wide range of other social and political collective action in civil society take place,[10] but even where ecological issues are

[8] Smitu Kothari, "Social Movements and the Redefinition of Democracy," in *India Briefing* (Boulder, Colo.: Westview Press, 1993), pp. 131–62; Pramod Parajuli, "Prelude to People-Nation: Developmentalist State, Nationality, and Citizenship in India," in *Nationality and Citizenship: Global View*, ed. T. K. Ooman (New Delhi: Sage Publications, 1994); Pramod Parajuli, "Power and Knowledge in Development Discourse: New Social Movements and the State in India," *International Social Science Journal* 127 (1991): 173–90; Arturo Escobar and Sonia Alvarez, eds., *The Making of Social Movements in Latin America: Identity, Strategy, and Democracy* (Boulder, Colo.: Westview Press, 1992).

[9] Charles A. Reilly, "NGO Policy Makers and the Social Ecology of Development," *Grassroots Development* 17, no. 1 (1993): 25–35.

[10] The literature on social movements is vast. In addition to Escobar and Alvarez, see, for example, Gail Omvedt, *Reinventing Revolution: New Social Movements and the Socialist Tradition in India* (Armonk, N.Y.: M. E. Sharpe, 1994). For a comprehensive overview of social movements in the Western world, see Sidney Tarrow, *Power in Movement: Social Movements, Collective Action, and Politics* (Cambridge: Cambridge University Press, 1994); Alain Touraine, *The Return of the Actor: Social Theory in*

raised, there are often serious contradictions between movements—for instance, individual lifestyles of activists being at variance with ecological principles. In any case, in decades to come, the challenge will be to better understand and interpret these popular stirrings, which are increasingly providing a comprehensive ecological critique of the developmentalist state; articulating a basic critique of the institutions that are exercising global economic hegemony; linking movements for ecological justice with the memory of historical struggles;[11] opening up political space for governance and autonomy at the level of communities and within civil society, essentially as counter spaces to the domain of state and global capital; and articulating a creative fusion of modern and indigenous traditions of knowledge.

THE POVERTY OF THE DISCOURSE

In their ecological critique of the current patterns of economic development or in their assertion of newer identities based on an attachment to place or even in their basic disputes with the developmentalist state, these movements are challenging the entire range of "received ideologies." They question the so-called sustainable development people, who believe that the industrial mode of development can be sustained along with enlightened environmental policies, that is, that the challenge is to green growth toward a sustainable green capitalism. They reject the idea that modern science and technology are intrinsically superior, taking support from the critics of modern science and technology who have opened up a significant space for a discourse on the multiple traditions of science and technology, a process that coprivileges other forms and systems of knowledge.

These movements also question those naturalists and conservationists who propagate the preservation of biodiversity and biological reserves while being indifferent and even hostile to the people residing within or in the vicinity of these parks and sanctuaries. In contrast, advocates of cultural diversity (including those among tribal and nomadic communities) argue that identity (including an identification with place) and ecology can be fused into a politics of the future. They argue that despite political commitment and state funding, none of the

Postindustrial Society (Minneapolis: University of Minnesota Press, 1988); and Alberto Melucci, Nomads of the Present: Social Movements and Individual Needs in Contemporary Society (Philadelphia: Temple University Press, 1989).

[11] Richard Grove, "Colonial Conservation, Ecological Hegemony, and Popular Resistance: Towards a Global Synthesis," in Imperialism and the Natural World, ed. John MacKenzie (Manchester: Manchester University Press, 1990).

major actors—forest departments, national park administrations, conservationists, or global institutions—can save biodiversity without directly involving local communities and granting primary rights to those resources to them.[12]

Still, there continues to be a significant disjunction between most of this theorizing and research and the insights and reflections within the movements, many of which are not only opposing developmental projects, but also asserting a new politics of autonomy. Movements have critiqued this disjunction by arguing that most of the "articulators" have rarely engaged themselves with the complexities and trials of the struggles at the base of society, where there are growing examples of political mobilization by movements that are seriously pursuing efforts to define alternative systems of decentralized governance.

In countries like India, social movements reflect a diverse array of influences. They have drawn freely from indigenously evolved socialist and Gandhian perspectives, as well as the residual history of centuries of resistance against oppressors and colonizers. Other influences include the Marxist, the Marxist-Leninist, and elements of the Western Green movement. Given this complex and often contradictory legacy, as well as the larger global context of ideological drift, a more coherent ideological redefinition is still at a nascent stage. Despite this, movements have identified many elements that are contributing to the ongoing process of conceptual crystallization. The need to delineate more rigorously the root causes of ecological degradation (as well as the dominant cultural, economic, and political assumptions about nature), the resultant social disruption as well as species extinction, and the growing realization that there are serious limits to the "pursuit of progress" all underscore the serious limitations of contemporary political theory. Some commentators call for a shift from predominantly anthropocentric worldviews to ecocentric ones.

Most historical and anthropological studies of ecological movements fall short of capturing this dynamic, primarily because they confine themselves to the analysis of only one mode of production at a time (hunting-gathering, nomadic, tribal, settled agricultural, or industrial). Except for a few scholars,[13] these studies have not shown the

---

[12] Ashish Kothari, Saloni Suri, and Neena Singh, "Conservation in India: A New Direction," *Economic and Political Weekly*, October 28, 1995; Nancy Lee Peluso, *Rich Forests, Poor People: Resource Control and Resistance in Java* (Berkeley: University of California Press, 1994); Ramachandra Guha, "Towards a Cross-Cultural Environmental Ethic," *Alternatives* 15 (1990): 431–47; Smitu Kothari and Pramod Parajuli, "No Nature without Social Justice," in *Global Ecology*, ed. Wolfgang Sachs (London: Zed Books, 1993).

[13] Madhav Gadgil, "Deforestation: Problems and Prospects," *The Indian Journal of*

contradictions and conflicts between different coexisting modes that make possible an economy of exploitation. Madhav Gadgil and Ramachandra Guha provide an insightful analysis of the resource demands of the capitalist mode of production and its respective impact on pastoral, tribal, and agrarian modes.[14] In fact, most of the ecological contradictions emanate today because all the other modes of production are unevenly articulated (of course, in different degrees) with industrialism and the industrial economy. This is precisely why there is need for a deeper understanding, both of the nature of resource use within each mode of production and the power relations between the different modes, as well as how external factors (such as the information revolution or other processes of globalization) and internal changes alter the dynamics of these relationships.

In the rapidly changing political climate, ecological movements or social movements that are becoming more self-consciously ecological have begun to recognize new objects, new problems, and new values. There is thus the possibility of discursively constructing new antagonisms and forms of struggle, requiring, in turn, fresh analyses and new solutions. For example, many of these movements transcend the prevailing ideas that development is inevitable or that conservation is the answer to ecological problems. Instead, they show that ecological problems are the product both of dominant attitudes regarding the value of nature and of the uneven patterns of development between and within more industrialized and less industrialized economies. Since natural processes are essentially cyclical, reductionist methods of analysis in which nature is divided into slices are highly problematic. Obviously, ecological problems require political-ecological solutions. In short, *there will be little nature without justice and little justice without nature.*

It would be naive to conclude that social movements have to be the only actors in a transformational politics. The attempt here is not to privilege these movements exclusively but to recognize the diversity of "voices" that are represented and that provide a sustained critique of the dominant systems. We also need to acknowledge that there are differentiated voices within each movement, and that serious problems exist in the process of how "we" represent "them" or even how individual voices within represent what the movement "believes in" or

Public Administration 35, no. 3 (1989): 752–801; Madhav Gadgil and Ramachandra Guha, *This Fissured Land: An Ecological History of India* (New Delhi: Oxford University Press, 1992); Vandana Shiva, *Staying Alive: Women, Ecology, and Development* (London: Zed Books, 1988).
[14] Gadgil and Guha, *This Fissured Land.*

"stands for." Yet these struggles for ecological and social justice, often in alliance with one another and with supportive groups and individuals, do need to be privileged as the predominant and essential sources of challenge to the hegemonic interests of powerful political and economic actors.

ECOLOGY AS A CONTESTED TERRAIN

Antonio Gramsci's notion of hegemony can be invoked here to explicate the drama of power that is unfolding between state policies of development (increasingly directed by global capital and its supportive institutions) and these social and ecological stirrings.[15] Gramsci argued that class was not shaped by economic conditions alone, but has to be constructed by intellectuals as an agent of historical change. Extending Gramsci, it is even possible to argue that cultural factors, more than preexisting economic formations, are the basis of "class" and that a class or a group exercises hegemony over other classes not necessarily through the state, but through institutions in civil society—educational, religious, and cultural—to form a common social-moral language separate from, although interlinked with, the coercive domination of rulers. It is this hegemony that is being challenged by the sustained collective action of hitherto subjugated communities, who resemble a new class, one with a potent political force. It is important to note here that numerous social movements that are attempting to define a deeper ecological alternative cannot be categorized by class. For instance, anti-dam struggles like those in the Narmada valley or movements to protect the Amazon rain forest have participants that cut across the conventional economic categorization of class.

Thus several crucial questions need further investigation. Have the social and ecological movements referred to here reduced the hegemony of the national state by forming new spaces for local governance? Are these mini-transformations linked to other struggles of a similar nature? Within these movements has value-creation shifted from a market and profit orientation to relative local self-sufficiency and sustenance? To what extent are local traditions of science and technology becoming relegitimized? Have these movements evolved locally manageable, ecologically sustainable modes of economies and re-

---

[15] Antonio Gramsci, *Selections from Prison Notebooks*, ed. and trans. by Quinton Hoare and Geoffrey Nowell Smith (New York: International Publishers, 1971); Ramachandra Guha, *The Unquiet Woods: A Century of Protest in the Indian Himalaya* (New Delhi: International Publishers, 1989).

source use? To what extent has the hitherto suppressed knowledge of women, of indigenous peoples, of other "rooted" communities been reasserted as viable and desirable? Some of these ethical, political, and pragmatic questions are examined below, along with three other issues that are becoming increasingly important for the movements: sovereignty, the state, and knowledge.

Sovereignty

Invoked in a wide variety of contexts, "national sovereignty" has been one of the most important weapons of the state against internal and external critiques of the adverse impacts of economic development. For example, the criticism by the movement in the Narmada valley in India of the dams being built on the river is construed by the government as being antinational.[16] Then why is the almost total acquiescence in societies like India to the policy dictates of the International Monetary Fund and the World Bank not perceived as a fundamental threat to national sovereignty? There is a related paradox here. Even though the disaffection against the policies of the multilateral development banks and their economic regime is widespread, it has not as yet coalesced into a potent political force. In this situation, who else but the state can act as a buffer (or arbiter) against predatory transnational capital? Is the challenge then to democratize the state or is it, as many movements are demanding, to create new "nations" that converge with distinct cultural, ecological, or political identities?

State

With respect to the modern national state, we can envision at least three possible options for these movements. One possible option is the anarchist strategy of rejecting the state and developing counter-authorities.[17] The second option is to try to reform the liberal democratic state by making it environmentally accountable through legislation and strict enforcement. The orientation of an overwhelming range of organizations in the pre- and post-Rio situation reflects such a view. The third option is to democratize the state. This can be achieved by using the symbolic discourse of the liberal democratic tradition in order to decentralize the state, so that there are more opportunities for

[16] William F. Fisher, ed., *Toward Sustainable Development: Struggling over India's Narmada River* (Armonk, N.Y.: M. E. Sharpe, 1995); *Lokayan Bulletin* (special issue on the Narmada Dam), 9, nos. 3/4 (1991).

[17] James O'Connor, "A Political Strategy for Ecology Movements," *Capitalism, Nature, Socialism* 9 (1992): 1–6.

citizens to take responsibility for (some of) its activities;[18] activate and democratize civil society by enlarging it to encompass issues of identity, social justice, and ecology; and pluralize the notion of a national state such that a diversity of state-nations and other autonomous formations can coexist.

In India as elsewhere, some of the ecological movements inspired by Gandhian and indigenous socialist philosophies of local self-governance and some of those articulating an "our rule in our villages" politics are arguing for the third option. In fact, they have demonstrated how greater political and social space is opened up in the process of democratizing the state. In practice, as numerous examples from both the Third and First Worlds suggest, this is extremely difficult. Most often, the state becomes repressive when demands for democratization and the devolution of power to hitherto disempowered communities and groups come in conflict with class and cultural interests that have historically succeeded in mobilizing the state to their own ends. As already discussed, the situation gets complicated further in the context of the increasingly pervasive role that the global market economy and international economic institutions play in restructuring national priorities in directions that, most often, are in conflict with the wider realization of ecological and social justice.

Ironically, processes of globalization have often given representatives of movements greater leverage to mobilize across national boundaries, making both the state and other international actors more accountable. But as some movement activists argue, these transnational networks at times deflect energy away from the more difficult task of political mobilization within the country, which alone can ensure the devolution of power. They also argue that most transnational networking has encouraged the state and international economic institutions to become more sophisticated in projecting a more acceptable social and environmental image, while their basic economic agenda remains largely unchanged.

Knowledge

Devising alternatives to the pervasive knowledge empire of development is a challenging task indeed. The primary orientation of the development industry in the last fifty years has been to delegitimate such possibilities. While the dominant development discourse might have accepted, at least marginally, other economic or political systems, on

---

[18] Michael Walzer, "The Civil Society Argument," in *Dimensions of Radical Democracy*, ed. Chantal Mouffe (London: Verso, 1992), pp. 89–107.

the question of knowledge it has left little room. Other traditions of knowledge, which were not part of the dominant industrial-scientific worldview, were slowly (and at times, brutally) rendered obsolete or irrelevant, or at best considered as a residual factor in the march of progress. At worst, other traditions of knowledge were cast as obscurantist or superstitious with no basis in a rational world. The survival and dynamism of subaltern traditions of knowledge is further weakened by the common tendency—even among those who work with/for them—of treating subaltern consciousness as one that is on the verge of being coopted or coerced by ruling ideas.[19] That is why it is probably easier to critique dominant ideologies than to regenerate subaltern consciousness.

Ideologically, modern systems of knowledge are characterized by universality and objectivity. On the other hand, nonmodern systems are spatially and locally embedded and most often reflect a unity between the subject and the object in their practice. By privileging nonmodern knowledge, the attempt is not to blindly debunk the modern. Rather, it is to highlight the need to understand the politics of knowledge and to reveal the domination of the modern systems of knowledge over all others. Tariq Banuri and Frederique Marglin also argue that

> unlike modern knowledge which bases its claim to superiority on the basis of universal validity, local knowledge is bound by time and space, by contextual and moral factors. More importantly, it cannot be separated from larger moral and normative ends. In order to make knowledge universally applicable and valid it is necessary to disembed it from a larger epistemic framework which ties it to normative and social ends. To value universally applicable technical knowledge over local knowledge in which technical know-how is inseparable from larger normative social and environmental contexts leads one necessarily to deny local communities the right to continuity.[20]

This is all the more relevant because the subalterns possess knowledge that resists totalization and has the potential of countering the hegemonic tendencies of dominant knowledge and power, while in the process producing other spaces of subaltern significance. Ethnoregional and ecological movements, the sustained resistance to large dams and power projects, to transnational control of biodiversity or

[19] Homi Bhabha, "DissemiNation: Time, Narrative, and the Margins of the Modern Nation," in *Nation and Narration*, ed. H. Bhabha (London: Routledge, 1990), pp. 312–13.

[20] Tariq Banuri and Frederique Appfel Marglin, *Who Will Save the Forests: Knowledge, Power, and Environmental Destruction* (London: Zed Books, 1993).

more fundamentally to World Bank–IMF impelled structural adjustment programs or movements, are part of this reality. The emergent
socio-ecological identity enriches resistance to the efforts of integration by the capitalist state by affirming the local, the regional, and the
ethnic. Simultaneously, there are growing efforts to link across national boundaries, not only to seek accountability of international institutions but also to build global countervailing power. Though at a
nascent stage and marginal to the dominant actors, these linkages are
already challenging the limitations of pursuing democratic change and
ecological sanity from a superstructural context. The limits of elite-
defined participation for enhancing democracy are painfully clear.

ECOLOGICAL DISCOURSE

One of the weaknesses of the current ecological discourse is its neglect
of the ecological intent and content of historical struggles for social
justice. Most historiography of peasant, labor, and women's movements, while highlighting the so-called rebellion and revolution processes, neglects the vast terrain of subaltern praxis that favors "the
everyday forms of resistance" over major forms of mobilization. If we
look beneath the surface, social movements are increasingly sites in
which social actors have struggled to constitute new identities based
on place (encompassing ecological as well as emotional dimensions) as
means to open democratic spaces for autonomous action. Social movements, in this sense, are not "dramatic events" beginning at one date
and culminating at another, but a continuous process that has its ebbs
and peaks and continuously reinvents itself in response to changing
external and internal conditions.

   For culturally and economically marginalized people, struggles for
biodiversity are also struggles for cultural diversity. For them, nature
is not merely a biological entity—a repository of forest, mineral, and
water. It plays complex, multiple roles in their lives and contributes to
their cultural identity. What is being defended is not pristine nature
but "social nature." As Susanna Hecht and Alexander Cockburn argue in the case of Amazonia, ecology is about land and people where
the practice of justice restructures the concept of nature.[21] The position
of the people as defenders derives not so much from the concept of
"nature under threat" as from an integral and integrated relationship
with the land, water, and forest as the fundamental basis for "their

---

[21] Susanna Hecht and Alexander Cockburn, *The Fate of the Forest: Developers, Destroyers, and Defenders of the Amazon* (New York: HarperCollins, 1989).

own elemental struggle to survive."[22] Nature for a vast majority of them is the basis for their cosmology.

It can be argued, therefore, that ecology, like culture, is always "constituted by conflict." Both will increasingly be battlegrounds for competing hegemonies. As David Harvey observes about contested spaces, in the sharpening struggles for justice, there will be a "permanent tension between the free appropriation of space for individual and social purposes, and the domination of space through private property, the state, and other forms of class and social power."[23]

Some social struggles by poor people can be understood as ecological struggles.[24] The issues that these movements espouse are not merely economic—about exploitation of natural resources and labor by uneven capitalist development. Some of them defend traditional methods of resource use and diverse traditions of knowledge associated with these methods (e.g., the Chipko movement in India and the Rubber Tappers' movement in Amazonia). Others such as Karnataka Rajya Raitha Sangha (a farmer's movement in Karnataka state in south India) have opposed gene theft by transnational companies such as Cargill. The Sangha aims to expel all transnational seed corporations from India, freeing farming communities to decide about production, storage, and sale of their seeds.[25] Others such as the Narmada Bachao Andolan (Movement to Save the Narmada) protest big dams—in the process they are not only challenging the dominant development discourse, but are also evolving alternatives to it. Still others protest projects like coastal railways and highways as well as policies that will irreversibly destroy ecological and cultural diversity. These are not to be misunderstood as Luddite movements but assertions that are challenging conventional patterns of economic development that exclude and devalue cultural and biological diversity and justice.

[22] Donna Haraway, "The Promises of Monsters: A Regenerative Politics for Inappropriate/d Others," in Cultural Studies, ed. Lawrence Grossberg, Cary Nelson, and Paula A. Treichler (New York: Routledge, 1992), p. 310; Lokayan Bulletin (special issue on survival), 3, nos. 4/5 (1985).

[23] David Harvey, The Condition of Post-Modernity (New York: Basil Blackwell, 1989), p. 254.

[24] Joan Martinez-Alier, "Ecology and the Poor: A Neglected Dimension of Latin American History," Journal of Latin American Studies 23, no. 3 (1990): 621–39.

[25] The systematic pillage of the gene-rich Third World by the gene-poor advanced industrial world and the vesting of these genetic resources within the control of the advanced states and transnational corporations is itself a powerful window to understand the almost total lack of moral and ethical principles in the privatization of genetic resources—a process that is biased heavily in favor of the economic and political priorities of advanced industrial countries. See Jack R. Kloppenburg Jr., First the Seed: The Political Economy of Plant Biotechnology (Cambridge: Cambridge University Press, 1988). For a more recent analysis, see David Goodman and Michael Redclift, Refashioning Nature: Food, Ecology, and Culture (New York: Routledge, 1991).

From State to Communities as Stewards of Ecology

Can we posit "community" as a possible alternative arena of political praxis? Partha Chatterjee has argued that although both the state and civil-social institutions have been enveloped within the narrative of capital, community continues to lead a subterranean, potentially subversive life within it.[26] *Communities still have immanent possibilities of being the site of resistance to capitalist recolonization.* We do not consider community an abstract concept, rather we consider communities as places in which groups of people bound by a common ecological and cultural heritage regenerate as well as critique their situations through various kinds of social movements. In India, for instance, there are at least half a dozen movements seeking the redrawing of internal boundaries (which were drawn on the basis of dominant language groups) based on cultural identity. Inevitably, there is no guarantee that even if this were to become a reality—as indeed it has in the case of the Bodo tribals in the north eastern state of Assam or the Gorkhas in the state of West Bengal—communities in their newly defined "nations" would not apply unsustainable ecological practices or be socially unjust. Of course, this invokes an age-old debate on what (and who) constitutes just and unjust. Many spokespersons in movements argue that the issues cannot be resolved by forcing one set of universal principles on diverse communities, but can only be resolved in an ongoing debate between those who are conversant with modern and universal principles of justice and those within the community that defend their specific cultural norms. Kymlicka presents this tension in his chapter in this volume and elsewhere, though he does not adequately acknowledge the severe difficulties faced by culturally bound communities surrounded by hostile interests as well as the overwhelming climate of righteousness exuded by most modern development practitioners "doing development" in their midst.[27]

The emphasis on community instead of civil society is a purposeful one. In many respects, civil society has turned out to be the "ideological cement in the restructuring of contemporary nation-states."[28] So rather than assume a homogenous civil society as an alternative arena from the national state, we see a need to view civil society from var-

[26] Partha Chatterjee, "A Response to Taylor's Modes of Civil Society," *Public Culture* 3, no. 1 (1990): 130.

[27] Will Kymlicka, *Liberalism, Community, and Culture* (Oxford: Oxford University Press, 1989).

[28] Orlando Fals-Borda, "Social Movements and Political Power in Latin America," in Escobar and Alvarez, *The Making of Social Movements*, p. 312.

ious sites of power. Recent feminist critics have insightfully shown how the prevalent notion of civil society is patriarchal and tied to the interests of the political and economic elite.[29] Civil society as construed today is not the space in which every citizen can hope to attain social justice. Carol Pateman, for example, demonstrates that the early modern discussion of civil society and the state always supposed the exclusion of women from civil society and their confinement to the privacy of the household. She attributes this exclusion of women to the "fraternal social contract" as found in the initial philosophies of Locke, Hobbes, and Hegel. All these represent the household as the proper domain of the socially and politically dependent wife.[30] In short, civil society is established after the image of the civilized, European male, individual property holder. Women and people without property were excluded from this construct. While this situation has greatly improved, civil society still remains a limiting factor where the national and global nexus of economic and political power continues to subvert and repress democratic processes. Can civil society be democratized, or do other social cohesions (like the community) have to be given primacy?

Like women, other subordinate groups also create their own spaces, and these are, as Nancy Fraser has accurately observed, "parallel discursive arenas where members of subordinate groups invent and circulate counterdiscourses, which in turn permit them to formulate oppositional interpretations of their identities, interests and needs."[31]

The notion of a community, then, needs to be invoked and constructed with great care, since there are multiple identities within it and it occupies multiple positions in the larger frame of domination and insubordination. For example, in general, indigenous people are marginalized vis-à-vis those who have come from outside, such as ranchers, farmers, contractors, bureaucrats, lawyers, and traders, and who reap the greatest share of the benefits of the development enterprise. Even inside communities there are social and gender hierarchies. The gulf, for example, between local people who have jobs in government or industry and those who are solely dependent on land, forests, and water is widening.

The growing recognition of ecological principles and of the increas-

[29] Carol Pateman, *The Disorder of Women: Democracy, Feminism, and Political Theory* (Stanford: Stanford University Press, 1989); Nancy Fraser, "Rethinking the Public Sphere: A Contribution to the Critique of Actually Existing Democracy," *Social Text* 25/26 (1990): 56–80.
[30] Pateman, *The Disorder of Women*.
[31] Fraser, "Rethinking the Public Sphere," p. 67.

ing self-confidence of communities that are asserting greater control over their resources is compelling a rethinking of conventional democratic processes and democratic institutions. The struggles for social justice need to be coupled with ecological justice—restoring rights to nature's resources to those who depend on them most as well as living within the limits of ecological sustainability. In a dialectical way, then, the notion of justice will have to redefine conventional political economy in ecological terms and ecology in political-economic terms. What will also need redefinition is the terrain of civil liberties and human rights, toward a wider body of democratic rights that will have to include the immensely difficult issues of the rights not just of future generations but of other living species as well.

Equally, significant rethinking is on the cards regarding the kinds of institutions that are meant to monitor the compliance of these rights and to ensure their enforcement. There are obvious problems in seeking a single institution to play global policeman, even if this role is in the interests of ecological justice. Can such a reconstruction be sensitive to and incorporate the survival, identity, and ecological claims from the grassroots? Can this reconstruction encompass not just issues of social equity but of intergenerational and interspecies responsibility?

Ecological movements and the efforts to redefine political theory toward an ecologically sustainable, socially just, postindustrial society thus requires a fundamental questioning—of existing political theory; of our attitudes toward Nature; of the role of the state; of the role of the market in the struggle for ecological and social justice; of the roles of community and civil society; of the politics of the "commons."

CONFLICTS OF INTEREST AND CONSCIENCE

As we move into the next century, we will be increasingly confronted with two conflicts: those of interests, which I have alluded to above, and those of conscience, the latter broadly encompassing struggles for social and ecological justice. We are poised at a very challenging crossroads. Does humankind as a whole have a right to the future? As so much evidence is now forcefully showing, we are losing our productive relationship with the planet itself. Among the many issues that have been highlighted by the social and ecological movements discussed in this chapter, four need emphasizing:

1. The need to transcend reductionist economics and the technocratic managerial ethic by recognizing the fact that entire peoples, ecosys-

tems, and nations have been rendered impoverished and destitute by the current patterns of economic development and ecological imperialism. Social equity is not a long-term project that can wait, and social justice is not feasible without ecological sustainability.

2. Justice and sustainability are also not feasible without confronting modes of consumption and patterns of production that legitimate and engender environmental degradation and widespread impoverishment. Current rates of consumption of the industrialized world and of Third World elites are incompatible with the realization of sustainability and democratic goals. Decrease supply, yes; decrease waste, yes; locate and use substitutes, yes. But what of the appetite itself?

3. While in the interim the processes of seeking accountability of states, national governments, and global institutions must continue, it is imperative that greater recognition, support, and legitimacy be accorded those movements, groups, institutions, and mechanisms that empower and facilitate relative self-sufficiency of local communities, nations, and the planet itself.

4. Multilateral institutions like the World Bank as well as transnational corporations and legal regimes like WTO, are, beyond a point, incapable of reform and beyond regulation. They are also undemocratic and unaccountable and inconsonant with the needs of a democratic and egalitarian order.[32] Again, while both strategies of holding them accountable as well as weakening their hold on the global and national economics must continue, the search for more accountable, more transparent, and more accessible and democratic institutions must be a priority. Movements assert that in the absence of this recognition, commitments to democracy ring shallow. They argue that if we demand basic democratic values in our interpersonal lives, we must

[32] For a critique of the World Bank that shows how institutionally incapable it is of promoting ecological and ethical values, see Bruce Rich, *Mortgaging the Earth: The World Bank, Environmental Impoverishment, and the Crisis of Development* (Boston: Beacon Press, 1994). For country case studies, see Bade Onimode, *The IMF, the World Bank, and African Debt: The Social and Political Impact* (London: Zed Books and the Institute for African Alternatives, 1989), and Robin Broad, *Unequal Alliance: The World Bank, the IMF, and the Philippines* (Berkeley: University of California Press, 1990). On the growing power and lack of accountability of transnational corporations, see Richard J. Barnet and John Cavanagh, *Global Dreams: Imperial Corporations and the New World Order* (New York: Simon & Schuster, 1994), and David C. Korten, *When Corporations Rule the World* (West Hartford, Conn.: Kumarian Press, 1995). For a related critical analysis of these institutions in the context of the United Nations, see Smitu Kothari, "Where Are the People? The United Nations, Global Economic Institutions, and Governance" in *The United Nations: Between Sovereignty and Global Governance?*, eds. Christian Reus-Smit, Anthony Jarvis, and Alberto Paolini (London: Macmillan, 1996).

demand it of the institutions that are defining how the planet will be organized.

It is obvious that ecological politics emphasizes that the political struggle toward an emancipatory future is going to be a massive one, one that as Henry Shue argues in Chapter 1 in this book, is full of "political obstacles," and one that will require the creative alliance of a host of diverse actors. He also rightly assesses the difficulty in reconciling the legitimate demands of economic development with ecological justice. Yet there seem to be few other options but to align with the alternatives that at countless sites all over the world are being practiced and expanded. The struggle calls for nothing short of reversing the conquest of society by the economy; of breaking down gross inequities within and between nations; of restoring the ethic of self-limitation; of recognizing that positive values are not just material outcomes but dignity, an integration of an ecological politics and a respect for the space for political struggle. Real development is not feasible without listening to our conscience, without restoring an ethics to economics, an ethics to development, without restoring to environmentalism an ethics and an aesthetics, without confronting dominant power and building countervailing power, without being fundamentally responsible in what we do to the most neglected child on this planet.

8

Science and Norms in Global Environmental Regimes

SHEILA JASANOFF

SCIENTIFIC knowledge enjoys a privileged place in current discussions of international relations because of the apparently boundless power of science to explain phenomena and to offer solutions. With the growing saliency of issues such as hunger, disease, environmental decay, and international security, the world community appears increasingly to have pinned its hopes for the future on the accumulation of technical information. Experts play an ever more influential role in defining and controlling fundamental social problems. Not only are their knowledge and know-how deemed essential for managing our most pressing problems, but science, because of its claims to value-neutrality, seems to provide the only forum where nations can set aside their differences in favor of a common, rationalistic approach to problem solving. To "scientize" an issue is at once to assert that there are systematic, discoverable methods for coping with it and to suggest that these approaches can be worked out independently of national or sectarian interests. Science represents for many the only universal discourse available in a multiply fragmented world. The yearning to resolve social problems by appealing to the apolitical authority of science persists, overriding the evidence that scientific knowledge is deeply embedded in politics and culture.

The perception that science can unite people from divergent perspectives and worldviews has gained support from studies of global environmental problems. Some attribute the very emergence of a *global* consciousness to the ability of modern science to elucidate previously uncomprehended facts about the functioning of the natural world.[1] The recognition of limits on the biosphere, in particular, is credited with speeding up the processes of transnational cooperation, as reflected in the proliferation of environmental agreements since the 1960s. Writing about the success of the ozone depletion treaties, the chief negotiator for the United States characteristically offered as the first and foremost reason "the indispensable *role of science*."[2] Many see the creation and dissemination of scientific information as among the most important contributions that international institutions can make to multilateral cooperation.[3] Others point to the increasing visibility of technical experts in national and international policymaking and argue that state action, especially on uncertain matters of environmental risk, is being shaped or even preempted by networks of knowledge-sharing professionals, usually termed "epistemic communities."[4]

The normative implications of this dependence on science have received relatively little critical attention from students of environmental change, although the essays in this book testify to the level of interest that moral philosophers and political theorists are beginning to take in other questions related to globalization. There is a tacit assumption that moral issues come to the fore only when science fails to generate agreements founded on a shared understanding of reality. As long as problems can be addressed "on the basis of science," the Western tradition seems prepared to hold in abeyance the concerns about equity and power that ordinarily attach to political decision making. Agreement among scientists thus has the effect of removing areas of discourse from ethical and political contestation. The literature on the policymaking role of transnational scientific communities, for instance, seems almost complacent about entrusting power to such knowledge elites. It is hard not to read as the subtext for these studies the post-Enlightenment faith in scientific progress and the concomitant view that increased knowledge will inevitably produce better social policy. The widespread belief that *scientific* knowledge can only advance the cause of global *environmental* policy seems but an unproblematic instantiation of the more pervasive modernist dogma.

[1] See, for example, Lynton K. Caldwell, *International Environmental Policy* (Durham, N.C.: Duke University Press, 1990), p. 9: "An enlarging of perception has occurred in which science has played an indispensable role."
[2] Richard E. Benedick, *Ozone Diplomacy: New Directions in Safeguarding the Planet* (Cambridge: Harvard University Press, 1991), p. 5.
[3] Peter M. Haas, Robert O. Keohane, and Marc A. Levy, eds., *Institutions for the Earth: Sources of Effective International Environmental Protection* (Cambridge: MIT Press, 1994).
[4] See, for instance, Peter M. Haas, ed., "Knowledge, Power, and International Policy Coordination," *International Organization* (special issue) 46 (1992).

At the same time, opinion polls, the news media, and popular culture convey the message that trust in the authority of science can no longer be taken for granted. The atomic bomb, the headlong buildup of nuclear arsenals by the superpowers during the cold war, radiation experiments and radioactive waste sites, and the spate of environmental and technological disasters in economically disadvantaged regions of the world—all these have contributed to a public perception of science and its attendant technologies that is as fearful as it is deferential. As we near the end of the century, there is a persistent intellectual unease about the directions and uses of modern scientific inquiry, about the wisdom and integrity of the inquirers, and, most particularly, about the dangers that may follow from an overly intimate alliance between science and the state.

The strain of skepticism deepens if we turn to recent academic studies of science and technology. Historical and sociological research over the last two decades has raised doubts about the separateness of science from society, and about the basis for its superior cognitive authority, by presenting accounts that stress the social origins of scientific claims. Scientific knowledge, according to a growing body of work, is a social construct in the sense that it is contingent on human interactions and susceptible to the multiple influences of economics, ideology, culture, and political interest. Indeed, the processes of science often resemble those of politics, and the similarities become most apparent when scientific activity draws near to the political arena through channels such as expert advice. In the limiting case, the scientific controversies that so frequently erupt in the course of policymaking simply retrace the patterns of more enduring conflicts in society.

To students of ethical issues in environmental policy, the socially embedded status of science poses a series of critical questions. Does the social constructivist account of scientific knowledge require us to reject science as a uniquely effective avenue for harmonizing disparate conceptions of the rights of human beings, their place in nature, and their obligations toward one another and their planet? How, in particular, can we reconcile contemporary views about the relativism of science with the continuing reliance of political actors on scientific solutions to social problems? What role should international scientific networks play in resolving policy controversies, and what are the normative implications of using science as the common language of global environmental politics?

To address these questions, it is necessary to reproblematize the variable of "scientific knowledge" so as to integrate it more fruitfully into theoretical discussions of global environmental policy. I would

like in this way to set the stage for a wider exploratory dialogue
among literatures that rarely intersect: social studies of science, political and moral theory, and global environmental studies. I hope to
show that how we characterize scientific knowledge in the international arena is not morally neutral but that it bears crucially on the
normative evaluation of agreements about the environment. To this
end, I begin with a brief historical fable that frames my central argument about knowledge and politics. I then review the strands of analysis in science and technology studies that have helped to illuminate
the cultural embeddedness of science and the normative and social
commitments that underlie claims of neutral scientific knowledge.
Using this critique, I propose seven different ways in which we can
deconstruct the work of the "epistemic communities" that are active
in the formation of international environmental regimes. I conclude by
reasserting the major political consequences that flow from this re-
framing of the role of science in the global environment.

A FABLE OF SUSTAINABILITY

In 1854, the naturalist and explorer Alfred Russel Wallace, codis-
coverer with Charles Darwin of the principle of natural selection, em-
barked on his celebrated journey across the islands of the Malay archipelago. A socialist and liberal thinker, Wallace developed a deep if
surprising admiration for the Dutch system of colonial government in
the East Indies, which he compared most favorably to Britain's con-
temporaneous rule over India. His account of the three-and-a-half
months he spent in Java, from July to October 1861, explains this
rather remarkable preference. It takes no more than a slightly ana-
chronistic stretch of the imagination for us to *re*vision what Wallace
"saw" in Java as the epitome of a sustainable society, where knowledge and power combined to maintain a harmonious balance between
human needs and nature. By visiting that world today, we more critical explorers from a later century can discern the constructed aspects
of the reality that seemed so convincing to Wallace. In this way, Wallace's adventure in Java becomes a key for deconstructing our own
contemporary myths of sustainability.

The crux of the arrangement that Wallace admired was the harness-
ing of an Eastern social structure to Western ideas of agricultural production. On the surface, the work of government was carried out by
native rulers, headed by princes or "regents"; but behind each regent
was "a Dutch Resident, or Assistant Resident, who is considered to be

his 'elder brother,' and whose 'orders' take the form of 'recommenda-tions,' which are however implicitly obeyed."[5] The residents (one is tempted to call them Dutch uncles) used the authority of the local rulers to induce their subjects to cultivate valuable cash crops that were sold to the government for a fixed price. A share of the profits was returned to the chiefs and the workers, with the result that, as Wallace observed, "on the whole, the people are well fed and decently clothed; and have acquired habits of scientific cultivation, which must be of service to them in the future."[6] With a scientist's discipline, Wal-lace applied an objective test—"the rate of increase of the popula-tion"—to assess this system's success. This measure led him to de-clare, with a fine Jane Austen-ian flourish, "It is universally admitted, that when a country increases rapidly in population, the people cannot be very greatly oppressed or very badly governed."[7]

The Javanese Eden that Wallace found so praiseworthy sustained itself, at one level, on the basis of formal knowledge (the precursors perhaps of modern Dutch agricultural science and welfare economics), but the science that regulated life in Java depended, as we see in hind-sight, on particular forms of moral and political order. The sus-tainability that Wallace observed could not have existed independently of established patterns of political and cognitive dominance, the for-mer embodied in despotic local rulers and their benignly invisible Dutch overlords, the latter in unchallenged presumptions of scientific cultivation and culturally grounded measures of the "good society." Is our own world, we may ask, any different in the way it integrates natural with social order? Are our own claims to scientific knowledge about sustainability any more universal than Wallace's?

THE SOCIAL CONSTRUCTION OF SCIENCE FOR POLICY

For many years, academic writing about science policy uncritically re-inforced the conventional wisdom that scientific knowledge could be created and validated independently of social influences, even if its primary function was to assist policy. Few analysts raised an eyebrow when senior officials at the U.S. Environmental Protection Agency (EPA) vowed "to get the science right" and to enhance the role of scientific peer review in carrying out the Bush administration's envi-

[5] Alfred Russel Wallace, *The Malay Archipelago* (New York: Dover, 1962), p. 73.
[6] Ibid., pp. 73-74.
[7] Ibid., p. 75.

ronmental program,[8] and no questions were asked when the new commissioner of the Food and Drug Administration turned to an expert
committee for recommendations on how to improve the flow of scientific advice into his agency. A more critical literature on science policy
is just beginning to take shape, but it has yet to penetrate the analytic
or political mainstream.

The first two decades of postwar collaboration between U.S. science
advisers and their government produced no deeper insight than the
well-known dictum that science can both act upon policy ("science in
policy") and be acted upon by it ("policy for science").[9] A 1970 essay
by the physicist and nuclear expert Alvin Weinberg was among the
earliest pieces to suggest that there might be unforeseen complexities
in the relationship between science and policy. Spurred by new and
controversial developments in the public management of risky technologies, Weinberg reflected on the implications of policymakers' increasing dependence on inadequate scientific evidence. Questions were
asked of science, he noted in a now famous passage, that science was
not in a position to answer; Weinberg dubbed these issues "trans-
science" and suggested that scientists had a special obligation to "inject discipline and order into the often chaotic trans-scientific debate."[10]
But although his essay underscored the problematic nature of the
boundary between science and policy, it adhered to the conventionally
separatist notion of the relationship between the two fields of en-
deavor. Science, as Weinberg saw it, was still something that could be
independently done by scientists and factored as needed into policy.
The only complication was that pure science was no longer sufficient
for solving social problems. Policymakers also needed to ask and resolve a variety of trans-scientific issues bearing on public decisions.

The scientific and technical controversies of the early 1970s, however, brought forth evidence that Weinberg's model was too simple to
capture all that was taking place at the nexus of science and policy. A
largely American body of case studies documented the important role
that political interests played in the production and interpretation of
science for policy purposes.[11] Whether the objective was to site a nuclear power plant, market a new drug or pesticide, or build a new
supersonic aircraft, it appeared that the proponents of technology con-

[8] Leslie Roberts, "Counting on Science at EPA," *Science* 249 (1990): 616–18.
[9] Harvey Brooks, "The Scientific Adviser," in *Scientists and National Policy-Making*,
ed. Robert Gilpin and Christopher Wright (New York: Columbia University Press,
1964), pp. 73–96.
[10] Alvin M. Weinberg, "Science and Trans-Science," *Minerva* 10 (1970): 220.
[11] See, for example, Dorothy Nelkin, ed., *Controversy: Politics of Technical Decisions*, 3d ed. (Newbury Park, Calif.: Sage, 1992).

sistently read the evidence of risk or safety differently from its opponents. As long as the science was uncertain—as Weinberg stated was always the case when predicting extremely improbable events—the likelihood of objective scientific analysis appeared to be extremely low. Organized interests were always on the alert to give the available information a spin that supported their own policy agendas.

While political scientists and policy analysts focused mostly on the subordination of science to interests, studies by cultural anthropologists argued that interests and preferences were themselves grounded in a social context, and that social organizations and institutions were therefore a more productive starting point for investigating how science was interpreted under conditions of uncertainty. Why, for example, did expert judgments about environmental degradation and deforestation in the Himalayas differ so radically on such seemingly scientific "facts" as "per capita fuelwood consumption"? Three researchers who addressed this question concluded that the experts' views were conditioned in every case by underlying socially induced perceptions of the problem. The "facts" that the experts had tried to discover about the Himalayas were so variable and context-bound, in short so irretrievably trans-scientific, that they could not possibly be ascertained through universal models of scientific inquiry. What the experts had to offer in the end were not descriptions of reality in any epistemologically tenable sense, but rather varying interpretations of uncertainty shaped by the assimilation of their methods and conclusions to their informants' local contexts and locally colored worldviews.[12]

Social organization (rather than shared factual knowledge) has also been offered as the explanation for divergences in the public perception of risks in contemporary technological societies. Building on a theoretical framework developed by the anthropologist Mary Douglas, several authors have argued, for instance, that risk perceptions are conditioned by two factors common to all organizations: the strength of a group's boundaries in relation to the rest of society, and the rigidity of the internal rules of prescription or differentiation within the group. Applying this model to the U.S. environmental movement, Douglas and her coauthor, the political scientist Aaron Wildavsky, suggested that entrepreneurial cultures, characterized as low on both the grouping and prescriptive dimensions, are generally most tolerant of the risks as well as the benefits of new technologies. By contrast,

[12] Michael Thompson, Michael Warburton, and T. Hatley, *Uncertainty on a Himalayan Scale* (London: Ethnographica, 1986).

Douglas and Wildavsky saw strongly grouped, egalitarian cultures as intrinsically risk-averse, because they are able to assert their groups' social identity most effectively by reacting against a perceived external threat.[13] Although the cultural theorists' limited explanatory framework remains controversial, the view that social and cultural arrangements shape public beliefs about hazards has won more general acceptance.[14]

If interests and cultural biases permeate the way people perceive risk, then traces of such bias can be expected to surface as different organizations and cultures analyze the scientific evidence purporting to establish that risks exist. Numerous cross-national studies of risk assessment have tended to confirm this hypothesis. In my own work on chemical regulation in Europe and the United States, which used political organization as the relevant explanatory variable, I found systematic differences in the way national regulatory experts interpreted evidence of carcinogenicity from animal and epidemiological studies.[15] Features of political culture, such as the extent and forms of public participation, influenced the reading of science, accounting in many cases for divergent interpretations of the same body of data. The open and adversarial style of U.S. regulation, in particular, focused attention on the uncertainties in the data, raised questions about the principles of risk assessment, required the development of "objective" assessment methodologies, and generally perpetuated technical debate in a manner that was virtually unknown in Europe at that time. The dynamics of scientific controversies as disclosed in these studies were remarkably similar to more general patterns of national regulatory politics.[16]

[13] Mary Douglas and Aaron Wildavsky, *Risk and Culture* (Berkeley: University of California Press, 1982). See also Michael Thompson, "To Hell with the Turkeys! A Diatribe Directed at the Pernicious Trepidity of the Current Intellectual Debate on Risk," in *Values at Risk*, ed. Douglas MacLean (Totowa, N.J.: Rowman & Allanheld, 1986), pp. 113–35. For an extension of cultural theory to the domain of individual risk perception and its empirical testing in that domain, see Karl Dake, "Myths of Nature: Culture and the Social Construction of Risk," *Journal of Social Issues* 48 (Winter 1992): 21–37.

[14] Brian Wynne, "Public Understanding of Science," in *Handbook of Science and Technology Studies*, ed. Sheila Jasanoff, Gerald E. Markle, James Petersen, and Trevor Pinch (Newbury Park, Calif.: Sage, 1995), pp. 361–88.

[15] See especially Sheila Jasanoff, *Risk Management and Political Culture* (New York: Russell Sage Foundation, 1986).

[16] See, for example, "The Politics of Scientific Uncertainty," chapter 8 in Ronald Brickman, Sheila Jasanoff, and Thomas Ilgen, *Controlling Chemicals: The Politics of Regulation in Europe and the United States* (Ithaca, N.Y.: Cornell University Press, 1985). Compare this account of scientific controversy with discussions of European and U.S. regulatory politics in such works as David Vogel, *National Styles of Regulation* (Ithaca, N.Y.: Cornell University Press, 1986); Joseph Badaracco Jr., *Loading the Dice*

Explanations based on social organization and institutional structure, however, are not sufficient in themselves to account for the myriad ways in which science splinters and kaleidoscopically rearranges itself in the policy context. Recent work in the sociology of science suggests that the possibility of divergent accounts of natural phenomena originates in the very means of production of scientific claims. Close observations of science in the making have revealed that the validity of scientific facts and theories depends on tacitly negotiated, often informal, interpretive conventions that are shared among communities of researchers.[17] Scientific claims, thus socially constructed, retain their authoritative status as long as their underlying premises are not too closely scrutinized by holders of different beliefs. They are, at the same time, always potentially open to deconstruction, particularly in the policy arena, where conflicts among competing interest groups and organizational cultures routinely impede consensus-building or closure around any particular account of "reality."[18]

At the boundary between science and policy, the definition of "science" itself is open to interpretation and negotiation. Weinberg partially grasped the possibility of such indeterminacy when he identified a bridging domain of trans-science between pure science and pure policy, but his analysis failed to recognize that assigning an issue to one or the other side of the science-policy boundary (or indeed to the intermediate domain of "trans-science") must always be a deeply political exercise. Labeling an issue as either "science" or "policy" implicitly entails an allocation of power—the power to speak or be heard on the issues in question—and interest groups will fight over these labels, as has been compellingly borne out by decades of controversy about issues such as the principles of cancer risk assessment in the United States. Regulatory agencies have stressed the *policy* content of these principles, in a move to preserve their discretionary right to articulate and modify them; industry and some segments of the scientific community, by contrast, have characterized the principles as *science* and pressed to assign their definition and revision to suitably qualified experts. While both sides recognize the importance of something called "scientific uncertainty," regulators construe this uncertainty as legitimating their policymaking role, whereas industry representatives see

---

(Boston: Harvard Business School Press, 1985); and Alan Irwin, *Risk and the Control of Technology* (Manchester: Manchester University Press, 1985).

[17] See, for example, Bruno Latour and Steve Woolgar, *Laboratory Life: The Construction of Scientific Facts* (Princeton: Princeton University Press, 1986).

[18] Sheila Jasanoff, *The Fifth Branch: Science Advisers as Policymakers* (Cambridge: Harvard University Press, 1990); also Jasanoff, *Risk Management and Political Culture*.

it as justification for the exercise of expert judgment. Similar forms of boundary drawing have occurred with respect to the definitions of "peer review" and "risk assessment," as parties with stakes in regulation have sought to emphasize or de-emphasize the "scientific" character of these procedures.[19]

Public controversies are the window through which we most commonly gain access to the negotiated boundaries of "science." Even in relatively closed and consensual decision-making cultures, controversies occasionally become public enough to force disclosure of the unarticulated, and generally unreflective, commitments about nature and society that lie unchallenged at the base of many supposedly scientific assessments. In the aftermath of Chernobyl, for example, British nuclear experts wrongly estimated the likely uptake of radioactive cesium by plants in northern sheepfarming country with acid, peaty soils. Their scientific models, based on studies in clay soils where cesium is chemically immobilized, had failed to take into account a key ecological variable—a mistake that only came to light when the measured levels of radioactivity exceeded the expert predictions.[20] The model in this case was based on a false presumption of ecological uniformity that might well have been disputed and set aside by the "nonexpert" sheepfarmers, who knew the variability of their own terrain, and on whose behalf the official model, ironically, was applied. Expert assessments are also known in many cases to incorporate untested and possibly unfounded assumptions about human behavior: for instance, risk assessors may routinely assume that workers will always wear protective clothing or that adults and children will consume equal quantities of food and drink, adjusted only by the factor of body weight. Unless criticism is institutionalized into the policy process at many levels, deference to the label "science" generally protects such unwarranted extensions of expert judgment from review by knowledgeable publics.

THE MEANING OF CLOSURE IN SCIENCE FOR POLICY

The subjectivity, relativism, and context-specificity of scientific claims lead us to be particularly attentive to the meaning of closure in policy-relevant science. Science, when it is not openly attached to policy, needs a certain level of containment in order to make progress. To be

[19] Sheila Jasanoff, "Contested Boundaries in Policy-Relevant Science," *Social Studies of Science* 17 (1987): 195–230. See also, Jasanoff, *Fifth Branch*, pp. 181–93.
[20] Brian Wynne, "Sheepfarming after Chernobyl," *Environment* 31 (1989): 11–15, 33–39.

successful, scientists have to accept the theories and claims of their predecessors as a point of departure for building other claims, which, in turn, can serve as building blocks for their successors. The historian of science Thomas Kuhn called this type of activity "normal science" and pointed out that it takes place within a governing "paradigm" that remains stable and frames inquiry for long periods of time but can be overthrown by scientific revolutions.[21] More recently, the French sociologist Bruno Latour has noted that scientists need to "black box" certain findings (that is, to shut them off against overly intrusive criticism) in order to propel forward the overall project of science, which is to build more and more "facts" about the real world.[22]

The scientist's primary rationale for closure (making more facts) ceases to be relevant, however, when science is engaged in policymaking. In this context, scientific facts function as a bridge to policy rather than to more facts, and the way is therefore opened for diverse interests to coopt the definition of salient facts in ways that advance their particular social and political leanings. As the terrain expands from normal science to normal politics, scientists, government officials, labor and industry, public interest groups, and even the news media all claim the right to place their own interpretive stamp on science. Advantage can be gained, moreover, not only by advancing one's own account of the "facts," but also by weakening the accounts of others. Thus procedures that allow for open criticism do indeed "bring new [scientific] data to light, and challenge gaps in reasoning," as one U.S. court foresaw,[23] but in the policy environment they also lead to protracted debate about the credibility of all scientific assertions.

Similar observations about the open-endedness of regulatory controversies led British policy analysts David Collingridge and Colin Reeve to argue that closure in science for policy could not be achieved without a prior closure in politics. The policy process, in their view, always presents an undercritical or an overcritical environment for scientific claims.[24] In the former case, there is already a strong consensus about policy and the addition of new science creates no new grounds for controversy. In the latter, policy adversaries who are separated by serious normative disagreements turn to science for help, but a socially constructed (and hence deconstructible) science is powerless to deliver

[21] Thomas Kuhn, *The Structure of Scientific Revolutions*, 2d ed. (Chicago: University of Chicago Press, 1970).

[22] Bruno Latour, *Science in Action* (Cambridge: Harvard University Press, 1987).

[23] David L. Bazelon, "Coping with Technology through the Legal Process," *Cornell Law Review* 62 (1977): 823.

[24] David Collingridge and Colin Reeve, *Science Speaks to Power* (New York: St. Martin's Press, 1986), pp. 28–34.

them from their political impasse. Instead of accepting science as a neutral arbiter, opposing interests subject each other's scientific evidence to heightened scrutiny, producing endless technical debate. Ironically, the transfer of value conflicts into the scientific arena undermines the authority of science by exposing its contingency and locally negotiated character even as it reaffirms science's power to mediate problematic social issues. This is the phenomenon that German sociologist Ulrich Beck refers to as "reflexive modernity."[25]

These types of claims accord quite well with the experiences of governmental agencies involved in science policymaking. Evidence from comparative policy shows that expert decisions carry greatest weight in precisely those environments where criticism is most muted or traditions of expert control are most entrenched—and where decision makers are correspondingly least likely to be pressured into defending the evidence they bring to bear on policy.[26] By contrast, closure is slow to form in U.S. regulatory politics, which demands an exceptionally high degree of transparency in the treatment of science, and hence is inhospitable to the boundedness (both epistemological and social) required for closing disputes. The regulation of carcinogenic chemicals at the EPA, as noted earlier, provides a case in point. Protracted debates over science and years of apparent stalemate on policy were commonplace in the EPA's attempts to set standards for chemicals suspected of causing cancer. Controversies about how to control carcinogens persisted more stubbornly in the United States than in other industrial countries sharing similar economic and public health concerns and confronted by similar gaps in knowledge. The comparative data merely underscore the intimate connections that exist in any society between the rules for forming political and epistemological closure.

THE GLOBAL ENVIRONMENT AND THE REEMERGENCE OF REALIST SCIENCE POLICY

What are the implications of the foregoing analysis for the international arena, where forces of economic and environmental interdependence are creating accelerating pressure for supranational approaches to science and technology policy? Neither academics nor decision makers have begun to address this question seriously, and the literature on international science policy accordingly lacks the theoretical

---

[25] Ulrich Beck, *Risikogesellschaft* (Frankfurt: Suhrkamp, 1986), translated under the title *Risk Society: Towards a New Modernity* (London: Sage, 1992).
[26] Brickman, Jasanoff, and Ilgen, *Controlling Chemicals*.

sophistication and empirical richness of studies focusing on national and comparative policy contexts. In particular, there is considerable work to be done to reconcile the account of policy-relevant science that I have presented with the positivist or realist strain evident in much current writing about the role of science in international environmental policy.

Historians of this century may well link the sudden political saliency of global environmental issues to the first satellite pictures of the Earth freely suspended in space, which offered powerful support for new ecological approaches to studying the interactions of humans with their environment. Here was visually irrefutable proof of the finiteness of the planet we inhabit, the interconnectedness of its living systems, and the futility of purely local solutions to global problems. The concept of the biosphere gained prominence, and the urge to understand it more completely led to authoritative advances in both the natural and social sciences. National funding priorities shifted toward the study of large systems or connections among systems, such as ecology, oceanography, climatology, chaos theory, and geography. Even the Gaia hypothesis won grudging respect from scientists who but a few years ago had dismissed it as the brainchild of a crank or mystic. Across the environmental sciences, there was a new emphasis on research that would bridge existing disciplinary divides and produce insights into the complex and ill-understood workings of planetary phenomena.

In policy analysis, as well, a new discourse emerged to encompass the unsettling image of Spaceship Earth, a discourse that tried to temper the freshly acquired sense of human limitations with hopes for a common scientific understanding and international cooperation. A seminal report from the World Commission on Environment and Development explicitly related the knowledge of planetary ecology to the possibility of intelligent human management.[27] In an influential discursive move, the Commission adopted the term "sustainable development" to signal the possibility of overcoming resource limitations through human enterprise. These two words seemed to sum up and yet transcend the contradictions of life and growth on a bounded planet. They suggested that the indefinite survival of the human spe-

[27] World Commission on Environment and Development, *Our Common Future* (Oxford: Oxford University Press, 1987). A specially telling passage was the following: "From space, we see a small and fragile ball dominated not by human activity and edifice but by a pattern of clouds, oceans, greenery, and soils. Humanity's inability to fit its activities into that pattern is changing planetary systems fundamentally. Many such changes are accompanied by life-threatening hazards, from environmental degradation to nuclear destruction. These new realities, from which there is no escape, must be recognized—and managed" (p. 308).

cies could be assured through a universally acceptable marriage between scientific knowledge and rational stewardship.

The Commission's report coincided with a renewed interest by social scientists in problems of global agenda setting and regime formation. As if reverting to a pre-constructivist and even pre-Weinberg model of science policy, in which knowledge unproblematically drives policy, many authors credited scientific knowledge with creating the momentum for worldwide cutbacks on the production and use of ozone-depleting chemicals,[28] as well as for multilateral discussions promoting wider consensus on marine pollution, greenhouse gases, biodiversity, and deforestation. Scientists and interpreters of science were identified as primary agents of policy change. According to an influential articulation of these views by political scientist Peter Haas, "epistemic communities" actuated by common professional norms and shared, holistic definitions of environmental problems were successfully setting the agenda for international negotiations and pressuring governments toward agreements.[29] Their efforts, Haas claimed, had in numerous instances persuaded states to overcome a reluctance based on rational utilitarian considerations and to adopt activist policies for environmental protection.

In this latter-day formulation, a new cadre of epistemologically authoritative high priests, equipped with the potent paradigm of ecological interconnectedness, are taking control of global environmental problems by "speaking truth to power" in benevolent harmony. The problem with this analysis, however, is that it leaves by the wayside virtually everything we know about the contingency and plurality of knowledge, the complexity of the interplay between knowledge and politics, and the immense efforts required to hold together global networks of technical practices and beliefs. In the next section, I propose an analytic framework that seeks to overcome these theoretical deficiencies. It also serves as the springboard for reintegrating the discussion of scientific agreements about the environment with normative and political analysis.

SEVEN TYPES OF AMBIGUITY

Epistemic communities, it is alleged, strengthen our theoretical framework for understanding why policy coordination occurs even in areas

---

[28] See, for example, Benedick, *Ozone Diplomacy*; Eugene B. Skolnikoff, *The Elusive Transformation* (Princeton: Princeton University Press, 1993).

[29] Peter M. Haas, "Do Regimes Matter? Epistemic Communities and Mediterranean Pollution Control," *International Organization* 43 (1989): 377–403. See also Haas, *Saving the Mediterranean* (New York: Columbia University Press, 1990).

of high scientific and political uncertainty, such as environmental protection.[30] But to understand how these communities achieve results, we must look more deeply at their own institutional characteristics. What are the features that enable them to overcome political resistance and act as agents of change? The factor of shared knowledge, the basis for the epistemic community's collective ideology, has received the greatest attention. As described by Haas and others, such communities need not be constituted exclusively of natural scientists, but can consist of "individuals from any discipline or profession who have a sufficiently strong claim to a body of knowledge that is valued by society."[31] Yet the commitments that are said to draw members of epistemic communities together clearly are fundamentally dependent on shared understandings of science. Four are usually listed in the literature: (1) a set of basic values that provide the rationale for social action; (2) a shared causal framework; (3) a set of validation criteria; and (4) a common policy project.

These generally accepted characterizations fall short of addressing a number of questions that would be seen as foundational from the standpoint of social studies of science; as I propose to show later, this failure is fraught with important normative consequences for environmental activists and policymakers. Why, to begin with, do epistemic communities coalesce into shared patterns of knowledge and belief? Put differently, what accounts for the magnetic power of the "episteme" that unites them? Is the policy project that epistemic communities share distinct from or an outgrowth of the common knowledge base to which they also jointly subscribe? More generally, what is the relationship between the belief patterns held by such communities and the knowledge claims produced by members of scientific disciplines or professions? Finally, what is the mechanism by which these groupings (or "natural coalitions")[32] achieve their cognitive authority in the political domain? The answers prove to be difficult, in part because of ambiguities in the way political scientists have used the term "epistemic" and in part because of theoretical concerns emanating from science studies.

We can abstract from the political and science studies literatures seven possible reasons why knowledge-based communities of believers

---

[30] I am indebted to Clark A. Miller for comments that led to the line of thought elaborated in this section. The section heading is borrowed with due respect and apologies from William Empson.

[31] Peter M. Haas, "Introduction: Epistemic Communities and International Policy Coordination," *International Organization* 46 (1992): 16.

[32] See James K. Sebenius, "Challenging Conventional Explanations of International Cooperation: Negotiation Analysis and the Case of Epistemic Communities," *International Organization* 46 (1992): 367–90.

might form or be created in response to environmental problems.[33] In the first four instances, the explanation for coalition building is "internal" in the sense that we look primarily at the nexus between the beliefs and the belief-holders; in the last three, the explanation turns "external" as we give primacy to the wider role of political and cultural forces in constituting epistemic communities.

Shared Factual Knowledge about Nature

Most writers about epistemic communities shy away from a simple positivism that postulates objective scientific "truths" as the glue that holds these coalitions together. Yet the uneasiness of the accommodation with social constructivist views of science surfaces at many points in the detailed narrative histories that these authors provide of environmental agreements. In writing about the Montreal Protocol on ozone depletion, for example, most accounts have accorded independent explanatory weight to the discovery of the Antarctic ozone hole and, more generally, to the formation of a scientific consensus on stratospheric ozone. Instead of treating this "discovery" as an object in need of social and political explanation (as it is conceived in the research program of the sociology of scientific knowledge), analysts such as Peter Haas and James Sebenius have externalized the ozone hole and advanced it as an independent factor empowering disparate communities of atmospheric scientists, economists, and engineers to influence policy.[34] Backed by poorly articulated models of science and technology and (sometimes) shallow empirical research, such attempts to explain international politics as a response to revealed nature have about them a strangely deus ex machina quality. More instructive are approaches that recognize the complex processes of cognitive, technical, and normative standardization that must operate together to build

[33] There is an interesting parallelism between the approach I propose here and that adopted by Richard K. Ashley in explaining why world modelers, a powerful epistemic community, have been so reluctant to engage in critical self-reflection. Ashley advances four possible reasons: a shared pragmatic assessment of priorities; the maturation of the field; common disciplinary affiliations; and a particular ideology of historical progress. See Richard K. Ashley, "The Eye of Power: The Politics of World Modeling," *International Organization* 37 (1983): 495–535. Each of his explanations is analogous to one of my seven ambiguities. Unlike Ashley, however, I see little a priori reason to give greater weight to one form of social explanation (e.g., ideology) over others (e.g., shared disciplinary commitments or professional interests).

[34] In this respect, the epistemic community literature hews surprisingly closely to the theoretically less sophisticated argument advanced by Richard Benedick in *Ozone Diplomacy*.

a robust consensus on environmental problems such as ozone depletion.[35]

Shared Causal Framework (or Paradigm)

Epistemic communities, in the standard account, share a causal paradigm that provides a common way of explaining problems in spite of uncertainty. Little has been said, however, about the origins of such shared worldviews, a fact that renders this part of the definition theoretically precarious. If we ascribe independent natural reality to the shared causal paradigms, we are tilted back in the direction of positivism. On the other hand, there are hints in the literature that common causal commitments may come about through social processes like shared disciplinary training or professional acculturation. Haas provides an example of such socially induced community building in the growth of a cadre of international experts who supported the Mediterranean Action Plan (MedPlan) and who participated to varying degrees in research and dissemination activities organized by the United Nations Environment Program.[36] Richard Ashley proposes the disciplinary affiliations of world modelers, most of whom were engineers, economists, or natural scientists, as a reason for the community's remarkable lack of self-reflection.[37] An explanation of epistemic communities that stresses this type of discipline-driven networking is quite consistent with social constructivism, and it cannot easily be disentangled from the "interests" explanations offered below.

Shared Disciplinary Interests

Standard accounts of epistemic communities tend to distinguish between the shared knowledge base and the shared policy commitments that bind such coalitions together. But numerous examples can be cited to show the tenuousness of this line of demarcation. I refer particularly to cases in which scientists committed to a knowledge paradigm have adopted a policy project that furthers their disciplinary

[35] See, for example, the account of the discovery of the ozone hole as a story about scientific instruments and standardization in Stephen C. Zehr, "Method, Scale, and Socio-Technical Networks: Problems of Standardization in Acid Rain, Ozone Depletion, and Global Warming," *Science Studies* 7 (1994): 47–58.

[36] These associations apparently included participation in UNEP-funded research and attendance at conferences sponsored by UNEP, as well as publication in UNEP manuals and a common core of professional journals. Haas, "Do Regimes Matter?" pp. 386–87.

[37] Ashley, "Eye of Power," p. 498.

standing and identity. The world modelers studied by Ashley provide one such example. In another, the Ecological Society of America adopted an influential public position on assessing the risk of releasing genetically engineered organisms into the environment. This action was prompted in large part by a desire to enhance the organization's professional standing; significantly, it postdated a report from the National Research Council, in which the institutionally more powerful community of molecular biologists and biochemists had articulated somewhat different principles, downplaying ecological consequences.[38] Other examples include the involvement of atmospheric scientists in promoting an activist policy on global warming during the hot summer of 1988[39] and the rejection of "clinical ecology" as scientifically invalid by groups of U.S. allergists and immunologists who were worried about the successful performance of this marginal field among prospective patients and in toxic injury lawsuits.[40]

Shared Commitment to Science

Enlarging on the foregoing point, we can plausibly speculate that scientifically trained members of epistemic communities will converge upon science as the most appropriate locus and discourse for resolving policy disputes about the environment. This, in effect, is a special case of the interests argument frequently advanced by sociologists of science to explain why closure occurs around particular causal interpretations of natural phenomena. Epistemic communities are often said to share a set of basic value commitments, including, presumably, the commitment to ecological explanations for environmental problems. Given a choice, they tend to define problems in ecological terms, although other possible framings may coexist in the social domain. To the extent that their shared "values" promote boundary-shifting (by

[38] For the position of the Ecological Society of America, see James M. Tiedje et al., "The Planned Introduction of Genetically Engineered Organisms: Ecological Considerations and Recommendations," *Ecology* 70 (1989): 298–315. For the view from molecular biology, see National Research Council, *Field Testing Genetically Modified Organisms: Framework for Decisions* (Washington, D.C.: National Academy Press, 1989).

[39] The attribution of political motives to climate scientists was easy enough in this case to have spawned a species of "backlash literature" by political conservatives opposed to international cooperation on global warming. See, in particular, Ronald Bailey, *Eco-Scam: The False Prophets of Ecological Apocalypse* (New York: St. Martin's Press, 1993); Robert C. Balling Jr., *The Heated Debate: Greenhouse Predictions Versus Climate Reality* (San Francisco: Pacific Research Institute for Public Policy, 1992); and Patrick J. Michaels, *Sound and Fury* (Washington, D.C.: Cato Institute, 1992).

[40] For a more detailed analysis of this case, see Sheila Jasanoff, "Judicial Construction of New Scientific Evidence," in *Critical Perspectives on Nonacademic Science and Engineering*, ed. Paul T. Durbin (Bethlehem, Pa.: Lehigh University Press, 1991), pp. 229–33.

allowing issues to be moved across the society-nature boundary from "politics" to "science"), it is tempting to interpret the values in question as nothing more than the overriding interest of scientists in enlarging the influence of science.

A historical parallel may be instructive here. Robert Merton's influential 1942 essay on the normative structure of science, together with its scholarly progeny, represented an earlier attempt by social scientists to ascribe a shared ethos to the natural sciences. Later scholarship has argued persuasively, however, that Merton's norms are better seen as a rhetorical resource that enhances the ability of scientists to claim a superior cognitive authority, and hence to widen their spheres of social and political influence.[41] It hardly needs belaboring that the scientization of growing numbers of policy arenas throughout this century— that is, defining society's problems in ways that call for resolution by experts—has enhanced the social, political, and (not insignificantly) material standing of the scientific community as a whole, even though some see in the proliferation of postmodern social identities, all seeking legitimation through science, a disintegration of science's post-Enlightenment hegemony.[42]

Whereas the four preceding deconstructions of the notion of epistemic communities took as the point of departure the communities themselves and their beliefs about nature and society, the next three focus on how such communities might be constituted or empowered through the action (or inaction) of the state and other forces in society.

Convergent Economic Interests of Business and Government

Epistemic communities have been held forth as a useful extension of rational choice models in explaining why international policy concordance occurs. Their activities are seen as instrumental in pushing decision makers, including nation-states, toward policy choices that would not be undertaken on the basis of coldly utilitarian considerations. In one frequently cited example, scientists at DuPont and other CFC manufacturing companies are said to have joined with epistemically

[41] Robert K. Merton, "The Normative Structure of Science" (1942), reprinted in Robert K. Merton and Norman W. Storer, *The Sociology of Science* (Chicago: University of Chicago Press, 1973), pp. 267–78. For a critical perspective on the norms, see particularly Michael J. Mulkay, "Norms and Ideology in Science," *Social Science Information* 15 (1976): 637–56.

[42] Yaron Ezrahi, "Science and Utopia in Late 20th Century Pluralist Democracy," in *Nineteen Eighty-Four: Science Between Utopia and Dystopia*, Sociology of the Sciences, no. 8, ed. Everett Mendelsohn and Helga Nowotny (Dordrecht, Netherlands: Reidel, 1984), pp. 273–90. See also Beck, *Risk Society*.

like-minded outside experts to argue for a reduction of CFCs, even though this policy was detrimental to their immediate institutional interests.

The science studies perspective teaches us to cast a critical eye on such stories of unexplained virtue. In the ozone depletion case, deeper historical research is surely needed to show why companies like Du-Pont, Germany's Hoechst, and eventually Britain's ICI dropped their opposition to international regulation and a CFC phaseout, but it is not hard to think of alternate explanations for this seemingly irrational outcome. First, however, we have to reconceptualize the question. Borrowing the language of negotiation analysis, let us say that the point needing explanation is why some CFC manufacturers decided to abandon the "blocking coalition" that they had previously formed with policy skeptics in their own national governments. This formulation, which focuses on why the nay-sayers receded rather than why the yea-sayers prevailed, leads us to a potentially very different reading of the ozone case. Our attention now turns on the U.S. side to such motivating factors in domestic politics as the draconian unilateral provisions of the U.S. Clean Air Act, the strength of U.S. public concern about CFCs, and industry sensitivity to the competitive disadvantage resulting from the absence of meaningful regulations in the European Community (EC). In Europe, we can point to the strength of the German Green movement as a factor in unblocking that country's, and eventually the EC's, previously monolithic opposition to CFC control.[43] Signing on to an international regime in the face of such political challenges could well have emerged as a rational choice for the CFC industries in policy-leader countries; for them, an international regime usefully leveled a playing field whose topography was in danger of being reconfigured by the vagaries of politics at home.

Convergent Domestic Policy Agendas

This explanation can be seen as a more general formulation of the preceding one. Instead of arising independently of, and in some sense prior to, the emergence of international accords, epistemic communities may develop out of the same complex constellation of forces that propel different state actors to adopt similar policy agendas. Again, detailed empirical work is needed to show just how "epistemic" and political convergence interpenetrate in particular case histories of policy development. I note, however, that the literature on

[43] Benedick provides an illuminating account of these country-by-country developments. See especially *Ozone Diplomacy*, pp. 30–32, 38–39, 84–85.

comparative social regulation includes numerous examples of governments arriving at rather similar policy outcomes through substantially different forms of domestic politics—and without necessarily agreeing on the "causes" of the problems they are addressing.[44] There is likewise considerable evidence that the traditional dynamics of interest group politics motivates nongovernmental organizations (NGOs) to disseminate scientific information favorable to their positions across national boundaries, creating pressure for policy convergence (see Steven Yearley's chapter in this volume on this point). In such cases there is little theoretical need to postulate a separate role for epistemic communities.

The story of the Montreal Protocol, in particular, could perhaps be told more convincingly as a study in the convergent evolution of national policies than as an instantiation of the theory of epistemic communities. While I can only sketch the argument here, it appears that the usual patterns of domestic politics in the United States and Germany were sufficient to account for the leading roles that these countries took in the effort to regulate CFCs, and even for the time lag between their respective actions. Thus all the predictable ingredients for a rapid policy response were present in the United States: the threat of cancer from ozone depletion, the consequent heightened public perception of risk, the adversarial relations between Congress and industry, with resultant ratcheting up of controls, the production of significant scientific knowledge and its opportunistic use by NGOs, and the presence of several mission-oriented agencies (EPA and the National Aeronautics and Space Administration, among others) to translate political demand into policy. Similarly, in Germany a corporatist decision-making style initially inhibited government action. Subsequently, the rising concerns of the Greens paved the way for a reappraisal of the risk and created a receptive environment for new scientific evidence on ozone depletion.

Shared Hegemonic Discourse

I come finally to what may be the most radical possible explanation for the rise of so-called epistemic communities on the global environmental stage: that far from fulfilling any independent agenda, these

[44] Thus, a study of chemical regulation in four industrialized countries showed that similar regulatory results were reached in all despite considerable differences in their legislative, administrative, and judicial processes. See Brickman, Jasanoff, and Ilgen, *Controlling Chemicals*. See also the studies of convergence on specific chemicals in Badaracco, *Loading the Dice* (vinyl chloride) and Jasanoff, *Risk Management* (formaldehyde).

coalitions are merely the instruments of a technological culture through which powerful industrial states impose a particular vision of natural and political order on the rest of the world. This line of argument takes its departure from work in the social sciences that calls attention to the coevolution of modern science and technology and the liberal state. Steven Shapin and Simon Schaffer's study of the scientific and philosophical controversy between Thomas Hobbes and Robert Boyle suggests that the rise of experimental science coincided with the structuring of a public political space with new rules of discourse and persuasion.[45] Yaron Ezrahi argues further that the modern experimental scientist's commitment to validation through witnessing by credible outsiders provided an indispensable legitimating technique for modern states and their instrumental social policies.[46] Writers on technology from Lewis Mumford to Langdon Winner have also drawn connections between politics and the material productions of technologically advanced states, noting particularly the incompatibility between certain large-scale technologies and the possibility of democratic control.[47] Seen in the light of such analyses, our "technoscience" (to use a currently fashionable term from science studies) and our politics are expressions of the same underlying public order; neither could be sustained without the other.

These critiques of science and technology lead one to question the easy proposition that epistemic communities, "based on shared cause-and-effect understandings, practices, and expectations," present an effective counterbalance to a hegemonically created world order.[48] To the extent that modern science both constitutes and is constituted by particular forms of politics, it can scarcely provide an independently authoritative mechanism for dealing with the destructive consequences of that political order, such as environmental degradation or ungovernable technologies. Sharers in a common scientific worldview are more likely to perpetuate than deeply challenge the political structures to which they are tied by bonds of reciprocal legitimation. This is precisely the conclusion that Ashley draws from his analysis of the technocratic world-modeling community.

[45] Steven Shapin and Simon Schaffer, *Leviathan and the Air-Pump: Hobbes, Boyle and the Experimental Life* (Princeton: Princeton University Press, 1985).
[46] Yaron Ezrahi, *The Descent of Icarus: Science and the Transformation of Contemporary Democracy* (Cambridge: Harvard University Press, 1990).
[47] See especially Langdon Winner, *The Whale and the Reactor* (Chicago: University of Chicago Press, 1986).
[48] Emanuel Adler and Peter M. Haas, "Conclusion: Epistemic Communities, World Order, and the Creation of a Reflective Research Program," *International Organization* 46 (1992): 389.

Evidence from the roster of recent environmental agreements provides some support for this rather negative appraisal of the role of science. When common understandings have been achieved in the international domain, it has more often been by virtue of ironing out relatively modest technical differences among participating states than by persuading them to reconsider seriously their moral stance in relation to nature. Thus converging scientific opinion about the risks of ozone depletion led in the end to a fairly traditional regulatory arrangement, focusing on cutbacks to production and consumption of the targeted pollutants and a planned introduction of safer alternatives. This "solution" followed the well-trodden path of seeking a technological fix for a technologically created problem; missing from the agenda of international debate was any serious consideration of alternatives that would require radical changes in the consumption patterns of the polluting countries.

Assent and Authority

A humane and enlightened Englishman visited the crown jewel of the Dutch East Indies in 1861 and found there the model of a sustainable society. What the Javanese who were thus intelligently governed thought of their condition is not recorded; by the scientific test of population increase Wallace surmised that they were happy. Today, we have the conceptual tools to be more reflexive about our own claims to knowledge and the moral and political commitments they may encapsulate. I have tried to show how such a critical gaze can be applied to the scientific productions underlying national and global environmental policies. In this concluding section, I point to some of the normative issues that come to light when the allegedly neutral contributions of transnational epistemic communities, in particular, are refracted through the skeptical lens of science studies. As we probe more deeply into the foundations of our "knowledge" of global phenomena, we expose a layer of politics and values often kept hidden by the mantle of science.

The first, and perhaps most important, issue is the nature and legitimacy of the authority that epistemic communities exercise in the international arena. The view from science studies suggests that—given the present highly imperfect state of knowledge about the global environment—the consensus represented by successful "epistemic" coalitions has to be sustained by deeply political commitments. There may be significantly different ways of accounting for the politics in specific cases. Thus epistemic communities may be formed through the self-interest of scientists, through a chance convergence of state interests, or through the macrodynamics of a hegemonic economic and political order. Once formed, however, the face that these coalitions present to the centers of political power is that of science or objective knowledge. They are often referred to as "knowledge-based" alliances. If they grow into "winning coalitions," their causal analyses of environmental phenomena are accorded a superior cognitive status. The implications of this potential discrepancy between the real and the represented bases for cognitive authority cry out for further empirical and normative exploration.

A second issue is the question of participation. Who actually participates in building the consensus represented by a successful epistemic community? The answer from social studies of science again does not encourage complacency. We know that in relatively closed political systems, where the opportunities for effective criticism are largely foreclosed, the contingency of knowledge can be overcome, and science can be invoked as an independent support for policy. Such policy environments are "undercritical" chiefly because they are underinclusive: they systematically exclude viewpoints that could lead to the deconstruction of dominant expert opinions. In open democratic environments, by contrast, criticism only begets more criticism, with policy stasis as the most likely result.

On the international scene as well, technically complex accords seem easiest to achieve when participation is initially narrowed. Analysts of the ozone depletion treaties largely agree, for example, that the incremental addition of countries—first the Toronto Group, then the European Community, only later China, and eventually India—was important to the success of the negotiations. But this progression also ensured the early exclusion of countries that were relatively ill prepared to participate in epistemic networks based on shared expertise, shared framings of environmental problems, and a shared understanding of the "causes" of ozone depletion. Not surprisingly, the agreements that resulted were quite conventional in their attribution of causality: in common with the majority of Western environmental laws, they targeted *emissions*, flights of inanimate particles from activities deemed benign in themselves, and not the lifestyles of the rich and famous.

What we exclude, finally, by accepting scientific knowledge—and its bearers, the new transnational epistemic communities—as independent analytic variables is the possibility of reviewing critically the questions of values and power that are always on the bargaining table of international environmental negotiations, whether the bargain is couched in epistemological or in political terms. How are some issues framed as "scientific" and who is doing the framing? What alternate problem definitions and policy approaches are discarded or disfavored? Whose knowledge counts as authoritative in matters of controversy, and why? What alternative systems of knowledge or belief are set aside or not even granted entry? These are the kinds of questions that should be put back at the center of a research agenda on the role of science in global environmental decision making. A skeptical second look at the knowledge claims of epistemic communities, together with a refocusing of attention on the normative foundations of international epistemic regimes, is one way to begin.

9

Campaigning and Critique: Public-Interest Groups and Environmental Change

STEVEN YEARLEY

IN the past twenty years the field of environmental policymaking has proved a fertile one for the expansion of science advising. All the countries of the North have, by now, institutionalized mechanisms for supplying expert advice on such issues as water quality, air pollution, countryside protection, and transport strategies. Indeed, environmental policymaking has turned out to be a particularly favorable arena for the operation of science advisers because many environmental problems have only been drawn to our attention in the first place through the specific cognitive apparatus of science. Without the endeavors of scientists, people would not have conceived of the ozone layer and certainly not of a hole developing within it. The anthropogenic greenhouse effect, too, is a source of danger to which only scientists could have been expected to alert us. In the late 1980s and early 1990s a great deal of attention was directed at the handling of large-scale international (in some cases putatively "global") problems, and once again scientific expertise came to the fore as policymakers sought to adopt positions on these problems. Supercomputers and super-big science, for example, are offered as the tools for understand-

The research on which this chapter is based was supported by a research grant from the Science Policy Support Group/Economic and Social Research Council (A09250006) and by a McCrea Research Award from the University of Ulster. I should like to express my thanks to Sheila Jasanoff and Judith Reppy for their extremely helpful guidance on the preparation of this chapter and to seminar audiences at the University of Lancaster, Queen's University of Belfast, Cornell University, Virginia Polytechnic Institute and State University, and the Science Museum in London, who raised many provocative points.

198

ing global warming or for managing various aspects of the global or international commons such as the Mediterranean, where pollution abatement has been monitored by an international group of scientists.[1] Science advisers are now gearing up for cross-national operations on a grand scale.

The role of the scientist in current environmental policy debates is far from straightforward. Certain environmental problems about which scientists advise can be seen to have arisen pretty much directly from the activities of scientists themselves: without biotechnology there would not be environmental concerns about the release of genetically modified organisms; without the huge programs of research and development on nuclear power there would be no anxieties about the disposal of radioactive waste.

Accordingly, while scientists have undoubtedly played a large role in the development of environmental policies, the position of the scientific community and of scientific advisers has been anything but unequivocal. To some degree, scientific expertise has come under attack simply for being inadequate; scientists have been seen to make mistakes or to radically reverse their views (for example over the allegedly problem-free character of nuclear power installations). On other occasions their shortcomings have been seen as tendentious; scientists are suspected of having given advice that suits their professional or commercial interests rather than that which corresponds with the public's good.

Such criticisms of science are to be understood in the context of a series of broader challenges to states' environmental policies, ranging from objections to road developments, through concerted opposition to nuclear power stations, to apparently unselfish campaigns for reform of whaling policies. The groups and individuals who framed these objections have themselves commonly employed scientific expertise, so much so that pressure groups and lobbyists have cultivated a body of counterexperts to confront the expertise available to the establishment. Such counterexperts have developed considerable skills in their specialist fields and have devised techniques for upstaging and discomforting the establishment's scientists.

So conspicuous have the experts of the antinuclear and public interest science movements become that the German sociologist Ulrich Beck has nominated environmental protest as a prime example of what he terms "reflexive modernization." Beck has won a wide audi-

[1] Peter M. Haas, "Do Regimes Matter? Epistemic Communities and Mediterranean Pollution Control," *International Organization* 43 (1989): 377–403.

ence for his claim that "postmodernism" should properly be understood as a crisis in modernist thinking precipitated by the application of modernist principles to themselves. For example, the more that fairness in recruitment practices is demanded and pursued through legal actions and the introduction of new sets of bureaucratic regulations, the more competing interpretations of fairness come to be exposed. The tireless pursuit of fairness ends up by tending to undermine the apparent clarity and universality of fairness itself—the very things that made it an attractive ideal in the first place. In this sense, according to Beck, postmodernism is better understood as reflexive modernism. In his view, the systematic application of critical analysis to science itself (in the philosophy and sociology of science for example) fits the reflexive pattern, since it identifies limitations in science's cognitive authority. As he puts it, "The *expansion* of science presupposes and conducts a *critique* of science and the existing practice of experts in a period when science concentrates on science."[2]

Beck is correct, I suggest, to identify an incipient paradox (or as he styles it a "crisis") here. The extensive deployment of scientific expertise and the close (as it were, "scientific") scrutiny to which it is subject are likely to reveal science's defects. As he expresses it:[3] "Science is involved in the origin and deepening of risk situations in civilization and a corresponding threefold crisis consciousness. Not only does the industrial utilization of scientific results create problems; science also provides the means—the categories and the cognitive equipment—required to recognize and present the problems as problems at all, or just not to do so. Finally, science also provides the prerequisites for 'overcoming' the threats for which it is responsible itself."

Yet it seems to me that there are two important limitations to the view he puts forward. First, Beck presents this "crisis" in the authority of science as though it were a problem principally for establishment science. But the proliferation of experts and counterexperts is every bit as much a problem for campaigning individuals and for pressure groups as it is for states. In fact, rather more so, since given the imbalance of power and authority between campaigners and official bodies, the difficulty has been more acute for environmentalists wishing to change the status quo than for the authorities and the establishment scientists seeking to maintain it. Public-interest groups have faced the challenge of arguing that current practices (of gaseous emissions or pesticide use or whatever) are dangerous and that official standards

[2] Ulrich Beck, *Risk Society* (London: Sage, 1992), p. 156.
[3] Ibid., p. 163.

are generally insufficiently strict. Commonly, it is they who have had to bear the burden of proof.

Second, as Sheila Jasanoff has aptly expressed it (in a personal communication), nongovernmental organizations (NGOs) have a need to be seen to be arguing "authentically." They generally have members whose interests they have to be seen to represent and from whose support they draw much of their political muscle. Reflexive modernity is not the only—perhaps not even the chief—problem they confront, since besides having to meet demands for technical correctness, they must secure legitimacy within the movement.

Aside from any theoretical interest in the situation in which NGOs find themselves, there are also good pragmatic grounds for examining the NGOs' approach to science, since NGOs are, increasingly, an influential social force. Such organizations impact in at least three ways on the response to the world's environmental problems. In the first place, they have played a big role in stimulating public awareness of and interest in environmental issues, in both industrialized and developing countries. Second, they have frequently acted as "watch dogs," aiming to hold governments and commercial organizations to their espoused standards and seeking publicity to shame those governments and companies perceived to be "bad performers." Third, and this is a rapidly developing trend, NGOs are moving from a primarily critical stance to one in which they offer policy proposals and aim to join with governments and commerce in the formulation of policies that affect the environment. In each role they use the skills and the rhetoric of science. Accordingly, in this chapter I shall examine how NGOs have handled the issue of scientific authority and how their attitudes to science and expertise have, in turn, affected their approach to campaigning in the international context. My focus on this development is intended to complement the analysis of science advising in an official or statutory context put forward in Jasanoff's chapter and in her recent publications.[4]

Converging Trends Toward Science among Green NGOs

One can cite various indices that show that the environmental movement's arguments depend on the authority of science, indeed increas-

[4] Sheila Jasanoff, *The Fifth Branch: Science Advisers as Policymakers* (Cambridge: Harvard University Press, 1990). For further discussion of science and the campaign work of NGOs, see Steven Yearley, *The Green Case: A Sociology of Environmental Issues, Arguments, and Politics* (London: Routledge, 1992), pp. 113–48.

ingly so. Of course, in some cases conservation and wildlife organizations were, from their nineteenth-century beginnings, heavily influenced by—indeed commonly overwhelmingly composed of—scientists. In many European countries, a leading aim of early conservation groups was the conservation of sites of value to ecological science. The rationale offered for systematic conservation was that it would facilitate the growth of scientific knowledge. In Britain this orientation is still strongly reflected in the leading habitat designation, the Site of Special Scientific Interest, whose declaration depends on satisfying criteria of scientific merit (rarity, typicality, and so on), rather than meeting, say, aesthetic or recreational standards.

In the North American context this orientation existed alongside an almost antithetical interest in the control and subjugation of nature and the elimination of threats to cultivation and civilization. As Donald Worster comments, "A major feature of the crusade for resource conservation was a deliberate campaign to destroy wild animals—one of the most efficient, well-organized, and well-financed such efforts in all of man's history." After World War I, scientists formed the principal opposition to this policy, arguing that it should not be pursued too far, since "the extermination of any species, predatory or not, in any faunal district, is a serious loss to science."[5] Thus, again, a potent argument used by voluntary conservationists—and in this case one used in diametrical opposition to state policy—concerns the value of nature for science's sake. In Continental Europe too, early nongovernmental moves for the protection of wildlife drew on the perceived interests of the scientific community in the preservation of the natural environment.

The nongovernmental organizations founded in the next major period of heightened social awareness of environmental issues, around 1970, were typically not dominated by scientists, at least not scientists concerned to campaign for environmental protection in the name of the furtherance of science. The distinction here must not be overstated: scientists were by no means absent from these campaigning groups. My point is illustrated in the story of the founding of Greenpeace: the originators of the campaign that set this organization on its feet included a radical Sierra Club member (Jim Bohlen) with a scientific occupation and a student of ecology (Patrick Moore), as well as a couple of lawyers and campaigning media personnel. Moreover, the occasion of the first protest action, opposition to nuclear testing on

---

[5] Donald Worster, *Nature's Economy: A History of Ecological Ideas* (Cambridge: Cambridge University Press, 1985), pp. 261 and 276.

Amchitka, a remote island at the end of the chain of Aleutian Islands, was justified in part as the protection of an island sanctuary of great importance to wildlife. But the scientists involved didn't want to protect the island or to end nuclear testing in order better to practice their science. Nor were the scientific skills of the personnel central to the development and prosecution of the campaign: Bohlen was employed as a "forest-products" scientist.[6] Similarly, in his study of environmental activists, Pearce reports an interview with Patrick Moore in which Moore, discussing Greenpeace's subsequent anti-whaling campaign, asserted that "the scientific debate about whether whales really are in danger of extinction is not one we want to get *reduced to*."[7]

In the ensuing twenty years, however, Greenpeace's staff has become far more precisely scientifically trained. By and large the explanation for this is a functional one: increasingly, the fight against governments' and industry's defenses of their policies on acid pollution, on nuclear safety, or on the sustainable levels of whale populations has demanded technical expertise. As the arguments have become more detailed and subtle, greater expertise in each component area has been called for. In many cases those Greenpeacers who do not approve of this stance have left to pursue direct campaigning outside the organization.[8] In the late 1980s Greenpeace U.K. appointed a head of science, Jeremy Leggett, with very strong academic credentials, and founded the Greenpeace International Science Unit at the University of London. These developments have resulted in more attention being paid to "universalism" in Greenpeace's technical pronouncements. Leggett has introduced a form of peer review for Greenpeace reports, using external referees, and the Science Unit, because it is stationary and on dry land, unlike the organization's other laboratories, can more easily receive external certification.[9]

This increase in the importance ascribed to scientific training and expertise is reflected in the development of other, similar organizations as well. In the celebratory publication to mark its twenty-first anniversary, U.K. Friends of the Earth (FoE) cites as the very first item a statement by Geoffrey Lean, then environment correspondent of the authoritative weekly newspaper *The Observer*, who, in 1992, wrote that "about fifteen years ago, someone told me that Friends of the

[6] Michael Brown and John May, *The Greenpeace Story* (London: Dorling Kindersley, 1989), pp. 7–13.
[7] Fred Pearce, *Green Warriors: The People and the Politics behind the Environmental Revolution* (London: Bodley Head, 1991), pp. 26–27, emphasis added.
[8] Ibid., pp. 27–32 for an example.
[9] Ibid., p. 40.

Earth's campaigners were likely to know more about their subjects than the relevant Government Minister. I didn't believe it. Since then I have found that this has usually been the case. Friends of the Earth has maintained its reputation as a reliable and indispensable source of information."[10] This quotation is followed by a further testimony, this time from the head of the United Kingdom's official pollution-monitoring body, Her Majesty's Inspectorate of Pollution, praising the organization's "technical dialogue." The fact that the group plays up the quality of its information and gives pride of place in its anniversary publication to boasts about its technical expertise presents an "institutional body language" very different from early publicity from the movement organizations. This difference is demonstrated by the findings of Donald Greenberg's work, carried out at the start of the 1980s, on FoE's contemporary media image.[11] According to his analysis, the stunts and other attention-grabbing devices used at the time were interpreted by policymakers as evidence of the unscientific and emotional character of FoE's arguments. Care is taken to present a different image today.

This dependence on science affects not only the ethos of the movement organizations but some of the practicalities of their actions. Scientific information-gathering places heavy demands on resources of time and finance. Once your reputation is linked to the quality of the information you put out, you cannot give instantaneous responses in the way that a less fastidious organization might. Similarly, the commitment leads to a loss of flexibility, since campaigning expertise has to be marshaled around a set number of themes and, therefore, new issues—for example, biotechnology or a sudden environmental incident such as an oil-tanker wreck—cannot suddenly be taken on board, however urgent or demanding they may seem to the media, the public, or one's supporters. While Pearce seems to accept Greenpeace's own assessment that "scientists remain on tap rather than on top," there is an important sense in which the growth of expertise and of facilities leads to an inertial influence over policy.[12] Expertise is a more dominant factor than is normally acknowledged.

In summary, one might have expected to detect large differences in the orientation toward science of two groups: on the one hand, radical

[10] Friends of the Earth, *Twenty-One Years of Friends of the Earth* (London: FoE, 1992), p. 2.

[11] Donald W. Greenberg, "Staging Media Events to Achieve Legitimacy: A Case Study of Britain's Friends of the Earth," *Political Communication and Persuasion* 2 (1985): 347–62.

[12] Pearce, *Green Warriors*, p. 40.

environmentalist groups and, on the other, the early generation of conservation-oriented organizations—the latter augmented by a collection of wildlife conservation groups that have sprung up to fill gaps in the natural historical coverage of the leading organizations (in the United Kingdom these gaps correspond to the work of bodies such as bat groups, Plantlife, and the Marine Conservation Society). While there are differences in their respective orientations, these are in fact less than would be anticipated because of the recognition of the importance of scientific and technical professionalism among the former and, of comparable importance but less relevance to my present argument, a growing acknowledgment of the essential role of campaigning among the latter. A similar convergence among U.S. organizations is noted by Barbara Bramble and Gareth Porter, though they focus less on the reasons for this trend. As they note, however, the U.S. situation is complicated by the existence of a "third type of environmental organization with . . . a small membership [which] derives its influence primarily from scientific-technical or legal expertise."[13] The World Resources Institute and the Environmental Defense Fund are cited as examples of this third type. Given that technical expertise is among their defining characteristics, it is clear that such organizations also fit into the pattern of convergence. Accordingly, there is increasing similarity among the majority of environmental groups in the North in relation to the importance they accord to science and the use of technically correct information. This allows me to turn, in the next section, to more or less common problems that the movement has with the use of science for effective campaigning.

IMPLICATIONS OF THE SCIENTIZING OF NGOs

Given these perfectly reasonable steps that have been taken toward technical sophistication on the part of leading NGOs in the industrialized world, we can ask what implications for campaigning and critique have followed. As already indicated, an important change has taken place in both ethos and personnel. In some cases, the ethos and practical orientation of NGOs can be strongly influenced by the predilections of their scientific employees: naturalist-dominated conservation bodies may still find themselves more inclined to field science,

[13] Barbara J. Bramble and Gareth Porter, "Non-governmental Organizations and the Making of US International Environmental Policy," in *The International Politics of the Environment*, ed. Andrew Hurrell and Benedict Kingsbury (Oxford: Clarendon Press, 1992), p. 318.

surveying, and reserve management—things that scientists actually *like* doing—than to campaigning, communications, and lobbying—activities with which scientists are less at ease. But apart from the kinds of activities on which they choose to spend their time, the increasingly scientific identity of environmental NGOs has had implications for the way that knowledge is treated. The difficulties they have faced reflect several of the points observed by Jasanoff in her studies of the role of expertise in the work of agencies such as the EPA.[14] But the position of NGOs' experts does differ from that of official scientists. In part this is because NGOs typically have access to fewer resources than official agencies. There are, moreover, two further considerations. First, as mentioned earlier, public-interest groups have to worry about winning and retaining public support; their actions must be seen as legitimate and should preferably be popular as well as technically justified. Second, as has also already been observed, the role of NGOs has primarily been a critical one (attacking and aiming to change existing standards). They have only recently been placed in the position of having to propose and defend specific policy alternatives. The difficulties they face can usefully be considered under two headings.

Scientific Authority

To examine NGOs' troubled relationship with the authoritativeness of scientific expertise, it will be helpful partially to reinstate the distinction between scientifically oriented conservation groups and campaign groups which I aimed to minimize earlier on. Conservation bodies such as the United Kingdom's Royal Society for Nature Conservation (RSNC) have a profoundly scientific identity and may well, as in the case of the RSNC, have had a long history of interaction with official scientists. Interviews with its members, study of its publications, and observation of its groups in action indicate that its staff holds a generally benevolent view of science.

To illustrate what I mean, let us take the example of public inquiries that I have observed during fieldwork in Northern Ireland. In two of these cases, the scientific evidence that the local branch of the RSNC either supplied or supported fared badly. In one case the scientific basis used by conservationists for classifying the value of habitats was attacked by a developer's legal representative.[15] Conservation scientists had devised a "scoring system" for evaluating competing sites that

---

[14] For example, Jasanoff, *Fifth Branch*, pp. 234–41.
[15] Steven Yearley, "Bog Standards: Science and Conservation at a Public Inquiry," *Social Studies of Science* 19 (1989): 421–38.

were candidates for declaration as protected areas. A developer was seeking planning permission for commercial operations on one of the high-scoring sites, and his legal representative skillfully undermined the scoring system. In the other case, the RSNC's local body appeared at an inquiry into competing plans for a marina development in a dual role: as an objector and, through its legally independent commercial arm, as a consultant to one of the developers.[16] The lawyer representing a rival developer sought to show that the conservation body could not be impartial, since its interests were tied to those of the company that was paying its consultancy fees. Part of its evidence was therefore discounted. In discussions during the inquiries and in subsequent interviews it was clear that in neither case did the conservation organization's representatives take this as indicative of a problem with the social "application" of scientific authority or with, as Beck would have it, reflexive modernization. Rather, they were inclined to interpret it as evidence of bias or clever manipulation on the part of lawyers. Science, they implied, was science, and if it was misconstrued that only showed how inadequately other institutions made objective judgments. For them, these problems do not reflect on difficulties endemic to the scientific enterprise.

On the face of it, groups whose primary orientation and identity is as campaign organizations tend to respond differently. Without the same scientific ethos and the same enthusiasm for scientific activity among their staff, they are freer to be critical of establishment science and of science as a form of cognition. But the move from an essentially critical stance to one that requires them to substitute their own analysis of the situation or to offer their own policy recommendations is likely to be problematic.

In some cases, potential difficulties may not become manifest, since the information necessary for positive campaigning can be uncontentiously derived. Thus in the celebrated case of (U.K.) FoE's popular and successful campaign against CFC-driven aerosols (in 1988–89) it was enough to argue that if U.S. authorities were convinced of the dangers to the ozone layer and if U.S. firms could withdraw CFCs from deodorants and other spray-can products, then British companies could follow suit.[17] The same argument was used in the late 1980s by Greenpeace U.K. against Ford's failure to fit catalytic converters or other emission control systems on its British models. Borrowing the

[16] Steven Yearley, "Skills, Deals, and Impartiality: The Sale of Environmental Consultancy Skills and Public Perceptions of Scientific Neutrality," *Social Studies of Science* 22 (1992): 435–53.

[17] Friends of the Earth, *The Aerosol Connection* (London: FoE, 1989).

company's own slogan ("Ford gives you more"), the Greenpeace campaign pointed out that a Ford car in Britain gives vastly more toxic pollution than one bought in the United States because of the company's response to the different pollution control regulations.[18] Essentially, in both cases the argument is one about double standards. If something is technically possible in one developed country, it must be equally possible in another. The argument does not need to appeal to any external indicators of technical correctness.

When it comes to contested scientific information these groups are in a less comfortable situation. Thus, on 12 August 1990 a television program was broadcast on Channel 4 casting doubt on the reality of global warming as a result of the enhanced greenhouse effect. It set out to question belief in warming and implied that climate scientists might be led into making exaggerated claims about climate change because the existence of such a threat would make it easy to acquire the resources for their research programs.[19] In the next issue of the FoE magazine *Earth Matters* a staff member attacked the broadcast by appealing to the weight of majority scientific opinion. The testimony of "about a dozen" dissident scientists had to be weighed against the views of the three-hundred or so scientists who wrote or reviewed the report to the Intergovernmental Panel on Climate Change (IPCC). The article went on to bolster this argument by noting that "Friends of the Earth has watched the scientific and political consensus grow in support of its calls to counter the serious risks from global warming."[20] No doubt it is quite reasonable for FoE's campaign staff to argue in this way. But the point arising from this example is that FoE has no scientific evidence of its own to use in settling this controversy. FoE's best argument is just that the most well-informed scientific opinion is on their side. They find themselves invoking the "consensus of scientific opinion" to overcome the TV journalists' deconstructive arguments—exactly the opposite of arguments they and other environmentalists have themselves deployed in other contexts: for example, against the official consensus on the supposed safety of nuclear plants.

It appears that two opposing attitudes to scientific authority can be adopted. Groups may embrace the supposedly disinterested and universalistic character of science; they are then ill prepared for cases when the authority of science is not robust under legal or political pressure. Alternatively, they can adopt a critical and skeptical ap-

---

[18] See, for example, *The Guardian*, 27 October 1988, 5.
[19] See the booklet prepared to accompany the program, *The Greenhouse Conspiracy* (London: Channel Four Television, 1990), p. 27.
[20] Andrew Dilworth, "Crisis or Conspiracy?" *Earth Matters* 9 (Autumn/Winter 1990): 4.

proach to science as a form of knowledge. But this leaves them in an interpretative fix when they want to support the judgments of expert panels or official scientific agencies, since they have only ad hoc ways of distinguishing between the science they support and the science they don't. Kevin Stairs, a Greenpeace lawyer, and Pete Taylor, a scientific adviser, acknowledge this problem indirectly. Describing the NGO experience of international lobbying and law-making, they cite an irritated delegate who complained to the others present: "It seems if I agree with you it is science, if I disagree with you it is policy!"[21] Their recounting of this remark appears calculated to show that they hold a sophisticated view of science, one which acknowledges that scientific and political judgments are never fully separable. To endorse such a view, however, is to store up problems for the occasion when Greenpeace wants to insist on the straightforward objectivity of scientific claims in some other domain.

Scientification in Campaigning

The problem of the fallibility or revocability of science is not the only one to emerge during the application of expertise to campaigning. In the case of FoE's response to controversy over global warming, environmentalists suffered from a kind of epistemological "feedback." The very tools they had used on previous occasions to throw doubt on expert opinion were now being employed to deconstruct views that they endorsed. They were besieged by their own rhetorical battalions.

But this is not the only kind of "bad feedback" that can arise from campaigners' use of science. For example, using scientific arguments to suggest that declining whale numbers threaten to put whole species at risk is an argument that offers cetacean hostages to fortune; as soon as evidence appears that numbers are picking up, it becomes difficult not to be seen as implicitly supporting the resumption of whaling. What this whale-numbers example demonstrates is that the turn to science isn't problematic only at an epistemological level. This move can also pose difficulties by encouraging campaigners to suppose that they can wholly substitute science for other kinds of consideration—in the case of whales, a moral concern.

Moreover, this problem does not arise only as a cognitive matter (in terms of the theories to which environmentalists lend credence); it has ramifications at the organizational and sociological levels too. I have

---

[21] Kevin Stairs and Peter Taylor, "Non-governmental Organizations and the Legal Protection of the Oceans: A Case Study," in *The International Politics of the Environment*, ed. Andrew Hurrell and Benedict Kingsbury (Oxford: Clarendon Press, 1992), p. 118.

already drawn attention to some examples of this. Thus the pursuit of scientific campaigning tends to cultivate staff with very specific expertise, who in part because of their respect for expertise, are reluctant to respond to issues about which they are not maximally informed. This point should not be exaggerated. It is clear that NGOs take an opportunistic approach to mounting arguments, just as governments and industries do; they are not straitjacketed by their commitment to expertise.[22] Still, they are reluctant to publicize arguments that they cannot back in detail and, the more expert they become, the higher their standards of acceptability rise.

Furthermore, expert campaign staff may also become distant from the lay membership whose environmental interests they are supposed to represent and on whose (financial and personal) support they depend. The possibility exists that the importance of scientific expertise has added to the distance between the professional cadres in leading social movement organizations and the casual movement supporter. While campaign professionals learn more and more about acid pollution or radioisotopes, "ordinary" members are encouraged to deal chiefly in terms of general principles such as the "polluter pays" or "precautionary" principles. Again, one should not try to make too much of this observation; many "ordinary" greens are very knowledgeable about ecological issues. Nonetheless, scientific specialization tends to inhibit rather than assist democratization in green campaigning.

Indeed, in Britain the groups that are both radical and effective tend to be centralized. Greenpeace is famously so, leading Robert Allen, a campaigning journalist working on waste incineration, to complain that "in Britain Greenpeace is very definitely bureaucratic and was seduced by the establishment fairly quickly. From a small grouping at the beginning of the eighties, Greenpeace displays all the trappings of a multinational company or a civil service department."[23] How rapid a seduction has to be to qualify as "fairly quick" is unclear to me, and—in any case—Greenpeace does continue to engage in direct actions and occasional lawbreaking, but Allen's apparent resentment stems from the alienation sometimes felt by community campaigners in the face of professionalized and wealthy campaign organizations. McCormick notes that through the 1980s Greenpeace became "less confrontational, and more inclined to use the same tactics of lobbying and discreet political influence once reserved by the more conservative groups."[24]

[22] This point was emphasized by Robin Grove-White of the University of Lancaster (personal communication).
[23] Robert Allen, Waste Not, Want Not (London: Earthscan, 1992), p. 223.
[24] John McCormick, British Politics and the Environment (London: Earthscan, 1991), p. 158.

In the United States, as the tone of Mark Dowie's recent book indicates, similar tensions have arisen. Dowie accuses movement organizations of having become too "polite." There is clearly a possibility for this tendency to be even more pronounced in the U.S. context where small-membership lobby groups have played such a conspicuous role in the shaping of environmentalism. Moreover, because such groups are often heavily dependent on support from foundations, there is a risk that they will avoid measures that threaten to alienate supporters in the foundations. Dowie charges that "mainstream environmental organizations are safe places for foundation philanthropy because they obey the unspoken dictum: 'Do nothing to jeopardize the value of your benefactor's endowment.'"[25] However correct this accusation, it is evident that well-funded small-membership organizations have different problems with legitimacy than mass-membership pressure groups do.

Friends of the Earth has self-consciously sought to develop an alternative model, institutionalizing an arrangement for cooperation between a centralized London-based staff and its regional local groups. The latter can select their own campaign targets but are bound by a license agreement with FoE that prevents them from acting in FoE's own name. Local groups have a form of shareholders' meeting at the annual conference but cannot require (only recommend) overall changes in policy. The headquarters staff have a highly professionalized ethos. The structure lessens but does not remove tensions between the core and the regionally active members. Even FoE, which has tried to build in a mechanism for sustained member participation, cannot be viewed as the *expression* of the movement or, for that matter, as its vanguard.

It is these twin issues of the authority of science and of the influence of scientification on campaigning and—indirectly—on representativeness that we need to bear in mind as we turn to look at NGOs' actions in the context of international campaigns.

As Experts onto the Global Campaign Stage

As I noted in the introduction, official science advisers are increasingly occupied with providing expertise in relation to international and, in some cases, putatively global ecological problems. NGOs are naturally following suit. But the ways in which they are making this move to

---

[25] Mark Dowie, *Losing Ground: American Environmentalism at the Close of the Twentieth Century* (Cambridge: MIT Press, 1995), p. 49.

supranational campaigning do not necessarily follow the same pattern as the establishment's experts.

If we take the case of acid precipitation in Europe, we find that the countries that first claimed to detect the effects (Scandinavian countries) were not the ones held to be primarily responsible; rather blame was directed at Britain with its many coal-fired power stations and prevailing westerly winds. The authorities in Britain long argued that the negative effects of acid rain were not proven and that even if acid rain were causing damage in Scandinavia, it was by no means certain that it was British acid rain that was to blame. Similar arguments divided Canada and the United States, though the North American situation was in many respects simpler than the European one, where the larger number of countries increased the range of possible suspects and multiplied the opportunities for passing the buck. In both cases the point is that, by and large, governments have an interest in contributing the minimum amount to the solution of problems experienced in common. Pressure groups, by contrast, follow a different logic. Claiming to represent the interests of the earth, of birds, of nature, or whatever, they argue that their governments should do better than simply not be the worst. They often press for their countries to take a lead or to join the best performers. The precautionary principle has also been widely invoked as an alternative methodological standard against which to assess interventions in the natural order. Thus there are reasons for thinking that green NGOs will take readily to addressing supranational problems—more readily than official national agencies will—and that their approach may well be distinctive.

Furthermore, environmental NGOs have not found it difficult in principle to ally their concerns with those of development campaigners. Jonathon Porritt, former director of FoE U.K. and associate of the Green Party, argued that the 1992 Earth Summit—derided at the time by nearly all environmental NGOs—did at least have one positive outcome in that it cemented in conservationists' minds the relationship between environmental and development campaign objectives. Increasingly, development NGOs and green groups are campaigning on similar themes. Environmental groups see that in certain campaign areas, green objectives are tied to development ones: in what is now the classic instance, logging is linked to the need for debt servicing. This fits easily with their acceptance that environmental problems are often intrinsically supranational.

At first sight, therefore, it appears that NGOs might find it relatively easy to go global; one might expect them to be able to live up to the injunction to act local and think global. Indeed, there is an additional

factor working in their favor. The international campaigning of movement groups has been assisted because various international bodies,
and in particular the United Nations and the European Union (EU),
have fastened onto environmental issues as a way to augment their
influence. The environment appeals to such bodies because they can
argue that it is inherently international and because it can be presented
as a public-interest issue. If, for example, the EU is putting forward
proposals that supposedly advance the public environmental good, it
is hard for national politicians to oppose these without seeming to
argue out of pure self-interest—seldom a rhetorician's favorite stance.
Analogous considerations apply to the case of the United Nations,
though it has much less say over governments' actions than the EU
does. Environmentalists may thus find these international bodies afford a "softer" lobbying target than do national governments. Still, it
should be noted that the costs and administrative demands of lobbying
at this level tend to screen out smaller environmental organizations
and thus to favor the larger campaign groups, thereby skewing the
campaign agenda toward those groups' concerns.

OBSTACLES TO EFFECTIVE GLOBALIZATION

Despite the favorable auguries, the problematic issues of scientific authority and of scientization reappear once pressure groups start to
campaign at a global level. In fact, in many respects the problems are
magnified.

First of all, ecological problems big enough to be considered global
tend to be technically complex. Being international, they also give rise
to conflicting interests. And these two factors interact, since there will
be many parties to any technical dispute. Consequently, the evidence
is likely to be contested and charges of tendentious interpretation are
likely to fly. The complexity of global climate modeling, for example,
means that while NGOs are well motivated to become involved, it is
hard for them to identify (and therefore occupy) the technical high
ground. Second, and worse still, the science involved is usually so
costly and time-consuming that the campaign groups could not expect
to take it on themselves to any great extent. Admittedly, with issues
such as global warming, the synthesis and review of existing information is itself an important scientific task and one in which Greenpeace,
for example, has sought to involve itself. All the same, NGOs are
bound to be dependent on science produced for other people's
agendas. Third, an option in national campaigning—pointing to the

benefits enjoyed by citizens in countries with more progressive policies—is not so readily available in the case of global problems. While
countries' policies on drinking-water standards or urban air pollution
can be set more or less independently, this is far less true in relation to
the protection of the ozone layer or other supranational phenomena—
various forms of "commons"—where the environmental impact on
the individual citizen may be only marginally affected by the actions of
their domestic government.

If these are the problems arising from attempting to deploy scientific
authority in an international context, allied difficulties arise from the
issue of the scientization of campaign issues. In their move to "global
thinking," northern NGOs have often assumed that environmental
"goods" to be sought after in the First World are presumably good for
the rest of the world too. But groups have run into controversy by
generalizing about the presumed interests of other peoples. And scientific expertise—and counterexpertise—have come to be central to
these disputes over the legitimacy of NGOs' stances.

Such difficulties are illustrated in the well-publicized and much-discussed case of the seal cull around northern Canada. In the 1970s and
through into the next decade, Greenpeace and the International Fund
for Animal Welfare (IFAW) campaigned to halt the killing of seals.
Indigenous hunters argued that they should be exempted from any
proposed ban, since they had "traditionally" hunted these animals,
which were essential for their customary lifestyle. Controversy subsequently centered on the extent to which seal culling was truly a traditional activity among indigenous peoples. These people now used
modern technology for seal hunting, and many campaigners chose to
interpret the use of motorized transport and guns as evidence of commercial hunting, particularly since such activity was, by definition,
nonsustainable without entering the cash nexus—in order to buy fuel,
ammunition, and snowmobiles. By contrast, community campaigners
and sympathetic anthropologists claimed that this was just a minor
adaptation of traditional practices. Given the changing population
structure of the Canadian Arctic it was only by the use of these modern technologies that a traditional way of life could be maintained at
all.[26] In this case both sides appealed to "scientific" assessments of
"traditional-ness"; if hunting were truly traditional then, presumably,
it might be allowed, whereas if it were demonstrably not traditional
then—like other commercial hunting—it too should be banned. But

[26] George Wenzel, *Animal Rights, Human Rights: Ecology, Economy, and Ideology in
the Canadian Arctic* (Toronto: University of Toronto Press, 1991), pp. 165–66.

rather than resolve the problem these appeals merely fueled the disagreement.

Similar difficulties arise, even more gravely, in relation to another
disputed question, population. For many in the North, particularly
those in right-leaning administrations, growing population levels (especially in the South) are self-evidently an environmental problem. Development campaigners from the South dispute this notion; they argue
that the North's excessive consumption is far more damaging to the
planet than the South's admittedly numerous but very low-consuming
citizens.

In relation to population the notable thing is the near-silence maintained by the North's leading NGOs. Green Parties, which are obliged
to have policies across the board, have at least been forced to confront
the question of what their ideal population level for, say, Germany or
the United Kingdom would be; they have also had to set out their
attitudes to immigration into Western Europe and so on. But the
North's environmental NGOs have very largely avoided campaigning
on this issue, even though it surfaces in all major international con-
frontations between policymakers of North and South—as, for example, at the Earth Summit and most conspicuously at the UN population conference in Cairo in 1994. Writing in *The Guardian*
immediately before the Cairo conference, Jonathon Porritt detected
"tight-lipped, catch-all caution," instead of which he recommended
that "the directors of Oxfam, Christian Aid, Friends of the Earth,
WWF and all the rest of them [bang] on Lynda Chalker's [the then
minister for overseas development] door to tell her that an absolute
priority for Britain's aid must be to help meet the needs of the 300
million women worldwide who want to limit their fertility but lack
access to any contraception."[27] This silence is explained, I suggest, by
two factors. First, NGOs' existing technical expertise has, to date, not
been concentrated in this area, since it has not been an issue for campaigning within the North. Thus inertia leads them to silence on an
issue of great practical importance for North-South environmental relations. Silence motivated by humility is compounded by a recognition
that the subject is highly contentious and that whatever view one
takes, one is likely to antagonize as many people as one attracts. Pragmatic judgments favor silence too.

In 1992 the FoE publication *Poverty, Population, and the Planet*
appeared.[28] In essence, this report adopts the analytic position I have

[27] Jonathon Porritt, "Birth of a New World Order," *The Guardian*, 2 September
1994, 8–9.
[28] Ronnie Taylor, *Poverty, Population, and the Planet* (London: FoE, 1992).

ascribed to development campaigners: that is, it casts the population problem as an issue of global overconsumption and thus redirects attention onto the North's "excessive" per capita consumption. This allows the report to suggest that FoE is already campaigning on the population problematic, since it is combating excessive consumption and pollution in the North. In this way, the "recommendations" section of the report has no need to offer any specific FoE policy or campaign around population levels.[29] Policy and activism on that issue is left to other (unnamed) NGOs and government. The group's adoption of such a stance is, I suggest, fully compatible with the analysis I offered of environmental NGOs' relative silence.

A further example of difficulties with the adequacy of technical expertise for campaigning on international themes comes from rainforest campaigns. Such campaigns indicate how technical expertise can be confounded by moral complexity. Rain forest protection has proved a great popular success in the North. Indeed, so much public interest has been excited by rain forest conservation that groups that previously had more or less nothing to do with the rain forests have begun to find distinctive ways into this publicity hot-spot. Among the cleverest such links is that made by Britain's largest wildlife organization, the Royal Society for the Protection of Birds, which identified an endangered rainforest bird on whose behalf it could campaign.[30] Given the recognized publicity benefits of such campaigning, it becomes important to justify this preoccupation on the part of northern NGOs, who have often been rather less emphatic in their concern to preserve northern forests or peat bogs, even though these too are major carbon reservoirs. And the exact rationale for campaigns focused on the South is open to being contested. Thus if the claim is that rain forest destruction contributes to the anthropogenic greenhouse gas, it has not been demonstrated that other campaign strategies (say, in favor of enhanced energy conservation) could not be more efficacious in limiting CO_2 accumulation. Northern groups' eagerness to conserve rain forests has also excited local mistrust and opposition because of the implicit threat to sovereignty and to people's rights to control over their own countries' resources.

Still, the final and perhaps starkest example of how northern NGOs' concerns with the supposed technical needs of the natural world can conflict with moral values comes from the World Wide

[29] Ibid., pp. 22–23.
[30] For further details, see Steven Yearley, "Standing In for Nature: The Practicalities of Environmental Organizations' Use of Science," in *Environmentalism: The View from Anthropology*, ed. Kay Milton (London: Routledge, 1993), pp. 59–72.

Fund for Nature's (WWF's) approach to wildlife conservation in Africa. At the end of the 1980s WWF was found to be assisting the Zimbabwean authorities with the purchase of a helicopter to assist in the protection of the black rhino, despite the authorities' known shoot-to-kill policy toward poachers.[31] The Fund also aided Kenya's wildlife rangers in purchasing assault weapons and helicopter gunships for hunting down elephant poachers.[32] Even leaving aside suggestions that national rivalries (offensives against Somalis) were being disguised as wildlife protection, this case demonstrates that very different standards are proposed in the Third World to those that would be accepted in the NGOs' northern homes. Again, when technical considerations are allowed to displace moral ones, some very contentious policies arise. WWF's actions, once publicized, generated an outcry in Britain.

A measure of support for the argument being developed here comes from the fact that some movement actors are beginning to acknowledge similar concerns. Thus it is notable that in the paper by Stairs and Taylor, the Greenpeace associates mentioned earlier, the authors comment:

> It is the opinion of the independent adviser in this joint authorship that NGOs need to take great care over their own internal dynamic with respect to motivation and policy. As NGOs grow to multinational million-dollar operations and are dependent upon public funding and media coverage of issues, certain inflexibilities and "hidden agendas" can develop. For example, Greenpeace evolved a policy of world-wide prohibition of sewage dumping very largely based upon the agenda of its North Sea campaigners. The transference of this policy to, for example coastal Malaysia, via the LDC [London Dumping Convention], without adequate research, could lead to environmental losses rather than gains—there, sewage is largely uncontaminated by industry, and coastal discharges threaten human health and wildlife: sea dumping could be justified if the cost of land treatment were unrealistic. On this issue, ethnocentrism, and campaign simplicity, led to an uncompromising stand.[33]

Though it is unclear from the wording of this passage whether the adoption of such a rigid universal policy "could" or did "lead to environmental losses," it is evident that some environmental movement actors now recognize that "universal" technical prescriptions may not be universally appropriate.

[31] Paul Brown, "WWF Paid for Helicopter Used to Kill Poachers," *The Guardian*, 4 September 1990, 1, 20.
[32] Pearce, *Green Warriors*, p. 75.
[33] Stairs and Taylor, "Non-governmental Organizations," pp. 133–34.

Globalizing from Limits to Growth

So far I have examined the difficulties arising when northern NGOs seek to move to a global level of campaigning by extension from their existing concerns. But an alternative approach employed by some is to work back from future limits rather than forward from current campaigns. This procedure is helpfully exemplified in a recent discussion document from Netherlands FoE (Milieu Defensie) called *Action Plan: Sustainable Netherlands*.[34] The basis for this document is the notion of "environmental space"—that is, access to the environmental resources needed to sustain a standard of welfare. Calculations of environmental space requirements were made under five headings: energy, water, nonrenewable resources (such as minerals), agricultural commodities, and timber/paper. These calculations were aimed at introducing sustainable exploitation by, in most cases, 2010, although it was accepted that the turnaround to energy sustainability would take twenty years longer.

In very many respects this piece of futurology is stimulating and original. It fully acknowledges the global nature of the environmental "problematique," emphasizing that "until now, the northern environmental movement has not really developed strategies towards a redistribution of the world's natural resources." Equally it departs from much environmental writing in taking very seriously the question of how the transition from the present state to the desired future state (sustainable and reasonably equitable demands on environmental resources) is to be accomplished. As the authors express it, their question is: "What will the consumption level be when we take into account the needs of all world citizens, and does this provide an attractive [prospect] for the average citizen in our affluent country?"[35]

Nonetheless, at a fundamental level the document is still highly technocratic. It doesn't deal with the kinds of government or participation that will allow for adherence to these agreements; it doesn't talk about the kinds of notions of worth and worthiness that will legitimize the distribution of resources within future society. There is no notion of the moral or the financial economy that will sustain the anticipated social order. The only explicit recognition of a political dimension comes during discussion of the distribution of environmental space, and the authors' response is to invite input from trade unions and other social-movement actors.

[34] Milieu Defensie, *Action Plan: Sustainable Netherlands* (Amsterdam: Milieu Defensie, 1992).
[35] Ibid., pp. 2 and 5.

The overall approach, a kind of limits-within-sustainability, seeks to attend in a sensitive way to the physical limits within which a global future must be planned. But without consideration of the socio-economic features of that future these calculations fall far short of a blueprint. Moreover, any projections of physical limits are themselves highly unlikely to be accepted consensually. Drawing on recent experience of other estimation exercises (projected fuel reserves, climate modeling, fish-stock management) we know both that there will be scope for legitimate differences of opinion and that people will have a tendentious interest in over- or underestimating the stocks and availability of the particular commodities in which they have a commercial or political interest. For the kinds of reasons outlined by Jasanoff in her analysis of science advising, no body of experts (however independent one tries to make them) could establish beyond doubt the "correct" values for these parameters. The Milieu Defensie authors do not confront or even really acknowledge this difficulty.

ENVIRONMENTAL NGOs of virtually all complexions are becoming more technically sophisticated. But despite this sophistication they face two sorts of problems: they are handicapped in attempts to "cash-in" on the authority of science and they face difficulties in avoiding the creeping domination of expertise and a technocratic outlook over their campaign style and strategy. Accordingly, far from being—in Beck's terms—the agents who bring reflexive modernization to light, these bodies themselves suffer from conflicts and contradictions in the use of expertise.

NGOs have good reasons for wishing to move to global campaigning, and at first sight the pragmatic conditions for globalizing their interests also appear favorable. But on reflection, the likelihood is that their problems in using expertise will persist on the global stage; indeed on balance the difficulties will tend to be enhanced rather than diminished. They are enhanced by the sheer complexity of international problems, by the lack of models of good practice that can be straightforwardly generalized to all countries (rich or poor, north or south), and by the ineradicable moral element in issues (such as "rights" to reproduce or to freedom from paramilitary nature conservationists) that affect environmental "management" in the south. This is true both in those cases where pressure groups extend their current campaigns into the global arena and where they start—so to speak—from the other end, by deducing what levels of consumption and pollution the entire planet can tolerate. In neither case does technical sophistication deliver all that might have been hoped.

This indicates, I suggest, that issues of equity and social justice cannot be omitted from their campaigning by the kind of "division of labor" apparently practiced by many NGOs and seemingly encouraged in Milieu Defensie's publication. As we have seen, there is some evidence that NGOs are examining organizational mechanisms for combating aspects of this problem. For example, in the run-up to the Earth Summit, Greenpeace representatives from the North went to some lengths to discuss their policy positions with representatives of southern NGOs in an effort to overcome ethnocentrism. Subsequently, attempts have been made to institutionalize these links in the Global Forum conferences for NGOs. But so long as northern NGOs formulate their policies toward international issues within boards composed exclusively of northern representatives answerable to a northern membership, they will be unable to claim a "global" democratic legitimacy. Since, as we have seen, technical expertise will not suffice to make good this democratic deficit, we can anticipate that this issue of ethnocentrism and representativeness will be at the heart of the politics of international environmental campaigning for years to come.

10

Breathing Room: Negotiations on Climate Change

PETER TIMMERMAN

AT the June 1992 Rio Earth Summit a long-gathering scientific debate about the possible global environmental consequences of human activities met an equally powerful and much more historically extended debate about the inequities of global development. The tensions and contradictions of that meeting were symbolized by the array of truncated documents that were the product of Rio: two wide-ranging protocols on climate change and biodiversity, based almost entirely on an uneasy mix of economic instruments and scientific research; a "forest principles" document of uncertain provenance; a giant "wish list" of tasks entitled *Agenda 21*; and a "Rio Declaration" made up of familiar UN pieties and an occasional fine turn of phrase.

The inability to fuse these documents into an integrated global vision or working plan is perhaps not surprising to those familiar with the workings of the United Nations system and with the historic tensions at work throughout the process. Yet part of the attractiveness of the "Earth Summit" was the sudden recognition that its timing, topics, and momentum were quite impressive; and that in such a context, if it could be sustained, the developing countries might actually have some substantial influence over the outcomes of the official efforts that had been set in motion. Perhaps even more important was the emergence of an outline of a global framework within which future maneuvers would have to take place. What made this framework more compelling than the usual rhetorical gestures about "one global community" was the sheer physicality at its heart: the image of the bounded sphere in space was actually beginning to affect, if not automatically to constrain, decision making.

This defining of human activity within an instantly recognizable "field" has, ironically enough, confused environmentalists as much as it has bureaucrats and politicians. In spite of years of talk about "thinking globally," the environmental community seems to have done little of this in any strategic sense. Nowhere is this more obvious than in the conflict over how to address the equity questions that arise over issues like climate warming or other aspects of global change. For example, since Rio, certain multinational corporations and national states have been using the concept of the "global commons" as a justification for engaging in "joint-implementation" agreements to cut back on global greenhouse-gas emissions. The environmental community cannot decide whether this is a new form of "eco-imperialism" or an appropriate commons tactic.

Since Rio, the struggle to relate environment and development issues to each other has moved into the detailed negotiating processes associated with the Rio conventions and, in particular, the funding mechanisms. These are much less visible processes and far less amenable to blunt forms of pressure from public opinion and nongovernmental organizations (NGOs). The momentary opening up of the debate to include fundamental questions about the equitable distribution of burdens and rewards in the context of global change has subsequently narrowed into battles over money. The political momentum for the larger strategic debate over the Earth has also dissipated or fragmented into a series of world conferences on related issues (population, women's issues, and so on), smaller institutions (e.g., the Commission on Sustainable Development of the United Nations), and various national efforts.

I believe that this confused situation derives from the inexperience of the international community in general with the return of nature (in the form of a finite planet) to the negotiating table. This is "Nature's Veto"[1]—the unsettling need to take into account the continued global coupling of human beings to physical systems after three hundred years of successful strategies of local uncoupling. The reassertion of the "pressure" of potential physical constraints on our activities goes against the grain of many social and political entities that have been able for so long to take the unfettered manipulation of the natural world as a given.

[1] Peter Harries-Jones, Abraham Rotstein, and Peter Timmerman, *Nature's Veto* (Toronto: Science for Peace, 1992).

Moreover, the climate change issue is one in which almost all the pressing facts are derived from physical systems models or extrapolated from past climates, which means that while the ultimate brute physicality of a warming is not in serious doubt,[2] almost all the details are. Or, to vary the "Nature's Veto" metaphor somewhat, it is like playing poker at a table where one player (Mother Earth behind the green eyeshades) has unspecified drawing rights for new cards.[3] Negotiators are presented with a situation where they are forced into some form of what has been called the "precautionary principle," without knowing how much precaution is required. The negotiations are further complicated by the disagreement between negotiators who subscribe to a belief that we are involved in a commons situation and those who do not, by the dispute over which strategy is most appropriate for a commons situation, and by a panoply of other priorities and concerns that vie for attention. Oran Young has recently characterized this as "integrative bargaining" where the uncertainty levels force a great deal of early exploration for mutually beneficial options, rather than staking out positions based solely on ascertainable self-interest.[4]

What I want to do here is bring some of the ethical issues involved in the contemporary debate over global environmental issues out into the open and assert their strategic relevance. While many of the recent debates and disputes at various conferences and negotiating sessions on global environmental agreements appear on the surface to be about such topics as technology transfer, carbon dioxide emission rates, debt-for-nature swaps, and so on, they invariably involve covert or overt reference to ethical claims and counterclaims. These claims have not, by and large, been handled very well by the various claimants.

---

[2] A recent summary of the evidence for the last one thousand years is Michael C. MacCracken, "The Evidence Mounts Up," *Nature* 376 (1995): 645–46.

[3] Mother Nature also seems to be holding all the "wild cards"—sudden flips from one stable climate state to another, raising the prospect of changes of up to ten degrees centigrade in a decade or two. Recent ice-core samples from Greenland suggest that the last interglacial period before our own was subject to instabilities of this type. See Greenland Ice-Core Project (GRIP) Members, "Climate Instability during the Last Interglacial Period Recorded in the GRIP Ice Core," *Nature* 364 (1993): 203–7. While the results are controversial, the underlying view that the climate system is capable of acting up in this way is becoming widely accepted. See W. S. Broccker, "Cooling the Tropics," *Nature* 376 (1995): 212–13, and P. C. Tzedakis, K. D. Bennett, and D. Magri, "Climate and the Pollen Record," *Nature* 370 (1994): 513. Such an emerging view would play havoc with the incremental management approach currently favored by policy-makers everywhere.

[4] Oran R. Young, "Negotiating an International Climate Regime: The Institutional Bargaining for Environmental Governance," in *Global Accord*, ed. Nazli Choucri (Cambridge: MIT Press, 1993), pp. 431–52.

DEVELOPING ARGUMENTS?

It is important to note, at the outset, that when global environmental issues first surfaced at an official global forum in 1972 (The Stockholm Conference), they were immediately connected—if only by refusal—to the North-South debate. It is only a slight exaggeration to suggest that at Stockholm environmental issues were considered by many developing countries at the time to be irrelevant to their immediate concerns—to be a "developed-country" issue. At Rio in 1992, remnants of this attitude could be heard in some of the language and positions taken by those nations that were at the margins of the two decades of environmental debate after Stockholm. This was partly because many of the bureaucrats and diplomats involved in the deliberations were unschooled in environmental issues (there were "bone-up" sessions prior to Rio) and tended to retreat into what was for them more familiar language.

One of the most important problems in the developed/developing country dialogue over the environment is the differential pace of the evolution of the environment as an issue. Although there are exceptions (especially among the most sophisticated NGOs), by and large the positioning of environmental issues in developing countries was (before Rio) almost identical to that of developed countries twenty years earlier. It was remarked to me by one U.S. negotiator before Rio that he was constantly sitting across the table listening to his earlier self—namely, that environmental issues were minor irritants, the price of progress, and a belching smokestack was a sign of prosperity. This is not to say that developing countries did not also perceive the issues differently: for one thing, poverty and the differential impacts of development were not basic elements in the formation of developed-country environmentalism (which also helps explain why Western environmentalism has so far failed to generate a workable social theory, in spite of the many connections between poverty and environmental deterioration in developed countries).

In many international environmental meetings between Stockholm and Rio, developing-country participants assumed or put forward, openly or surreptitiously, a standard array of factual and moral claims drawn from the development debate, which began over time to take on an environmental flavoring. Environment thus (at least at the international level of expression) was also cast within the standard developing/developed country conflict framework. This resulted in the following storyline, building on familiar quarrels:

Because the industrialized West is responsible for the skewed underdevelopment of the Third World, and the industrialized West is responsible for the largest share of global environmental damage (for example, carbon dioxide emissions), it should therefore bear the burden of its cleanup.[5] While the Third World has high population growth rates, this is more than offset by the high consumption rates of the North. The Third World must be allowed to develop in accordance with its own desires (even if these turn out to be versions of Western models of development). The Declaration of the Developing Countries on Climate Change that came out of the New Delhi Conference in 1989 repeats this story and gives priority to development over responding to climate change, if the two conflict.[6]

By and large, members of northern NGOs accepted this storyline, and the moral condemnations embedded in it, without demur. Developed-country government representatives and negotiators have tended, on the other hand, for a variety of fairly obvious reasons, to swerve away from this moral swamp and head for the safer and more familiar ground of high-level international bureaucratic pragmatism and horse-trading. To be cynical, the working assumption seems to be that sitting and listening to these accusations from developing countries is the rhetorical price one has to pay before getting down to the real questions, which "everyone who is anyone" knows are really about money.

THE FRAMEWORK CONVENTION ON CLIMATE CHANGE

With this kind of cynical eye, we might examine in more detail the Framework Convention on Climate Change (1992), which in its preamble notes "that the largest share of historical and current global emissions of greenhouse gases has originated in developed countries."[7] This is the closest that the Convention comes to acknowledging the developing-country version of what is called "carbon guilt." It is inter-

[5] The "history of development" debate is outside the provenance of this paper and the competence of the author. Among the myriad books see Bjorn Hettne, *Development Theory and the Three Worlds* (London: Longman Group, 1990).

[6] Developing country perspectives include "Perspectives from Developing Countries: Conference Statement," International Conference on Global Warming and Climate Change (New Delhi: Tata Energy Research Institute, 1989); "The Imperative of Equity: The Missing Dimension of UNCED" (statement of the South Asia NGO Summit, New Delhi, February 1992); and *For Earth's Sake: A Report from the Commission on Developing Countries and Global Change* (Ottawa: International Development Research Centre, 1992).

[7] *Report of the Intergovernmental Negotiating Committee for a Framework Convention on Climate Change* (A/AC.237/18 (Part II)/Add.1, 15 May 1992).

esting to note that in the negotiations over this phrase, India wanted the text to read "disproportionate current and historical share of emissions."[8]

The Framework Convention on Climate Change (FCCC) was adopted at the Rio Earth Summit, and following ratification by fifty countries came into effect on 21 March 1994.[9] The Conference of the Parties (COP) to the Convention met in Berlin in March–April 1995 and made little progress on what had been decided in 1992. Essentially, this convention is a general agreement, the details of which are to be fleshed out and agreed upon in a series of subsequent negotiations on specific protocols to the convention. These will depend very heavily on what is called the Intergovernmental Panel on Climate Change (IPCC), which is a consortium of international scientists and others who work on the scientific dimensions of the climate situation (Working Group I), on possible response strategies (the recently consolidated Working Group II), and on cross-cutting issues, for example, equity issues (Working Group III).[10]

The Framework is based on a negotiating model that was first used in the Montreal Protocol (1987) on ozone reduction.[11] This model was developed in contrast to (and out of the experience of) the Law of the Sea (1982).[12] One essential difference is that the Law of the Sea belonged to an older tradition of agreement, one that required very detailed understandings and agreements on a wide range of issues before signing and ratification could take place, and thus it took a great many years to be hammered out. The Montreal Protocol was designed to be a much more flexible instrument, and has—in spite of predictions that it would not work—been markedly successful.[13] Hampson notes: "What is both interesting and important about the ozone agreement is

[8] For further discussion and interpretation of the Convention, see James Cameron, "The Climate Change Convention: How It Was Made and What It Means," *International Journal of Environment and Pollution* 3, nos. 1–3 (1993): 67–77.

[9] A similar structure and process are in place for the Biodiversity Convention, which came into effect at the end of December 1993. The U.S. government, which had balked at signing the original convention, has agreed under the Clinton administration to sign, following "clarification" of aspects of the convention pertaining to intellectual property rights.

[10] For further information on the role of the IPCC in the policymaking process, see Bert Bolin, "Science and Policy Making," *Ambio* 23, no. 1 (1994): 25–39.

[11] Richard Elliot Benedick, *Ozone Diplomacy: New Directions in Safeguarding the Planet* (Cambridge: Harvard University Press, 1991); John Gribbin, *The Hole in the Sky* (New York: Bantam Books, 1988); Karen T. Litfin, *Ozone Discourses* (New York: Columbia University Press, 1994).

[12] Elisabeth Mann Borghese, *The Future of the Oceans: A Report to the Club of Rome* (Montreal: Harvest House, 1986).

[13] John Maddox, "Carbon Dioxide Will Be Harder," *Nature* 346 (1990): 11; Peter M. Morrisette, "The Montreal Protocol: Lessons for Formulating Policies for Global Warming," *Policy Studies Journal* 19 (Spring 1991): 152–61.

that although there are still many uncertainties . . . an international
regime has been created to co-ordinate international responses to the
problem of ozone depletion. The regime is sufficiently flexible to respond to new scientific information and has been devised in such a
way as to encourage more parties to become signatories."[14]

The Framework Climate Convention will operate within similar
mechanisms, which basically set down broadly agreed-upon parameters that are then focused and tightened when (or if) the situation dete-
riorates or new scientific information becomes available.

For this reason, like the Montreal Protocol, the Framework has
mechanisms to provide continuous upgrading and distribution of the
available scientific information, and it also provides grants and other
support to developing countries to limit their greenhouse-gas emissions. This funding mechanism includes the Global Environment Facility (GEF), which is an existing experimental fund in support of environmental projects. This fund will expedite the transfer of funds for
greenhouse-gas emissions reductions, to be administered (though this
is still controversial) by the World Bank, the United Nations Development Programme (UNDP), and the United Nations Environment Programme (UNEP), acting upon the advice of the COP.

The financial arrangements have been the most contentious issue
to date, though as will be discussed in a moment, the relation between the financial and the scientific issues is likely to loom much
larger as time passes. There were major debates at Rio (and after)
over the use of the Global Environment Facility, because of (1) its
association with the World Bank, and (2) the perception that the GEF
decision-making process is strongly weighted in favor of the donor
countries. Developing-country participants generally favored the creation of a new mechanism, which was not accepted. Nevertheless, because of Articles 11 and 21 of the Framework Climate Convention,
the GEF was to be "appropriately restructured" to have an "equitable
and balanced representation of all Parties within a transparent system
of governance."[15]

Questions were raised at the 1993 Beijing GEF meeting about existing GEF projects on biodiversity and coral reef protection, as well as
conflicts in its relationship with its parent World Bank.[16] Interestingly
enough, in response to issues raised by NGOs, the Malaysian ambas-
sador at this meeting noted, in support of the NGO concerns (includ-

[14] Fen O. Hampson, "Climate Change: Building International Coalitions of the Like-
Minded," *International Journal* 45 (Winter 1989–90): 36–74.
[15] The first phrase is from Article 21, referring back to the second phrase, from Article 11.
[16] Pratap Chatterjee, "In Search of Green Money: GEF Fest in Beijing," *Crosscurrents*
6 (June 1993): 6–7.

ing Friends of the Earth Malaysia): "You cannot address environmental issues unless you address questions of structural inequity. It is the entry point to sustainability." Although the GEF has been somewhat restructured, it has still only been retained for the Convention in an interim capacity.

The issue of what new climate-related projects would be funded under the GEF is contentious, as is the relationship between these projects and other unilateral initiatives outside the GEF. The Climate Change Convention makes reference to funding for covering "the full incremental costs" of actions taken to mitigate climate change, but (in part because of the conditions under which GEF funds are made available) there are questions about whether all national actions to limit emissions or sequester greenhouse gases could be funded, or only those with measurable "global benefit." Similarly, the negotiations have raised the issue of whether activities taken to adapt to possible climate change should be funded (such as building breakwaters around small island states), or only mitigation strategies. The very contentious question of whether or not "adaptation strategies" should be a priority at all remains open.

Determining what the criteria will be for acceptable funding through the GEF is likely to embroil the debate for years to come. As is well known, unlike the ozone debate, where the producers are fairly easy to locate, greenhouse-gas emissions and their mitigation embrace almost all social processes. Are only projects that directly reduce emissions to be funded, or are indirect projects that result in emissions reductions allowable? How indirect is too indirect? For example, would national industrial strategies come within the mandate of the fund?

The role of science in this debate is perhaps slightly more straightforward, but by no means simple. Discussions are under way about the establishment of a central data bank or inventory of the sources and sinks of greenhouse gases, which would be a small step forward. There are also discussions about funding for information exchanges (perhaps through UNEP). These are important steps, which were part of the Framework Convention package.

Nevertheless, an IPCC Report (Supplement 1992) notes soberly that there is a long way to go before even some of the most basic science is in place.[17] A brief sample of what is not known includes the following: The consensus models are still unable to account for something like

[17] Intergovernmental Panel on Climate Change, *Climate Change 1992: The Supplementary Report to the IPCC Scientific Assessment* (New York: Cambridge University Press, 1992).

one to two gigatons of carbon dioxide per year. There are many uncertainties in quantifying methane emissions from individual sources. The relationship between anthropogenic and natural sources of nitrous oxides and their effect on the increased abundance of atmospheric nitrous oxide is still unclear. And there is inadequate understanding for the purposes of modeling and prediction of: clouds, oceans, polar ice sheets, land-surface processes and feedbacks, and so on.

The ability of the developed country scientific establishment to reduce most of these uncertainties within the next five years to the point where clear policy responses could be directly related to the outcomes of modeling and predictive efforts is dubious. Moreover, regional impact studies from the climate models will continue to be imprecise into the foreseeable future, for some fundamental physical reasons.

Perhaps even more pertinent here is the role to be played by developing-country science. The developing-country capacity to be part of the scientific project, which is going to be instrumental in making the Framework Convention work, is best described by a paragraph in the Executive Summary of the IPCC Special Committee:

> The factors identified by the Special Committee as inhibiting the full participation of developing countries in the IPCC process are insufficient information, insufficient communication, limited human resources, institutional difficulties, and limited financial resources.[18]

The scientific uncertainties, then, will be matched or exacerbated by the formidable obstacles to decent inventorying, monitoring, and policing of local activities. That is, even if developing countries sign agreements, how can they ensure that these are lived up to, even with the best will in the world? The alternative prospect of an international monitoring or policing agency is already worrisome, for obvious reasons, to developing countries.

One of the reasons why developing countries have been reluctant partners in global agreements is that the lack of an adequate scientific presence in some countries acts as a drag on national policy formation on these issues. If we may borrow a phrase from the political philosopher, Jürgen Habermas, what we are not presented with here is an "ideal speech situation." Habermas's term refers to a discourse (or a negotiation) wherein all parties to the discourse can participate as equals. The lack of equal access to scientific information—and expertise—is one of the most powerful obstacles to an equitable process.[19]

[18] Ibid., p. 66.
[19] Jürgen Habermas, *Toward a Rational Society*, trans. J. Shapiro (Boston: Beacon Press, 1971), and *Moral Consciousness and Communicative Action*, trans. C. Lenhardt and S. W. Nicholson (Cambridge: MIT Press, 1990).

Also, apart from the need for developing-country scientists as "on the ground" assessors of compliance, the lack of such capacity impairs potential advances in the understanding of the mechanisms of greenhouse-gas sources and sinks in all parts of the tropics.

Last, there is the issue of what have already been referred to as "joint implementation strategies." These are experimental partnerships between developed and developing countries to reduce emissions in the developing countries essentially as a form of credit transfer. Norway has pioneered in this with a recent agreement with Mexico and Paraguay; the Netherlands has negotiated agreements with, among other countries, Malaysia. These partnerships, which include reforestation projects, technology transfer, and other activities, are likely to become a model approach for a number of countries. Developing-country NGOs are now referring to this process as "ecolonialism"—one developing-country delegate at a recent conference said to me: "The first time around they rearranged our country to remove the resources; this time they are going to rearrange it to put them back."

Nevertheless, there is now widespread interest in joint implementation[20] in both developed and developing countries, partly because the other proposed alternatives—a global carbon tax, global tradable permits, substantial transfers of new funds from North to South—are either stalled or implausible. One of the outcomes of the Berlin meeting in March 1995 was an agreement to initiate a pilot phase between developed and developing countries on a voluntary basis.

EQUITY, ECONOMICS, AND EARTH

To bring some order into the ethical issues raised by the complex situation described above, I borrow (and slightly adapt) a list from Henry Shue's chapter of points where questions of justice arise: (1) when allocating the costs of prevention; (2) when allocating the costs of coping (e.g., mitigation of impacts, and adaptation); (3) when determining the background allocation of resources for fair bargaining; and (4) when allocating emissions in the transition period, and as an ultimate goal. We can see from this list that the background allocation of

---

[20] A series of papers concerning developing-country perspectives on joint implementation that stress the benefits of supporting energy efficiency projects rather than carbon "sink" projects has been generated by the Indira Gandhi Institute for Development Research, Bombay, as part of the Indo-Canada Applied Economic and Business Policy Linkage Program, including Jyoti Parikh, "North-South Cooperation in Climate Change through Joint Implementation," 1994. Also, C. J. Jepma, ed., *The Feasibility of Joint Implementation* (Dordrecht: Kluwer Academic, 1995).

scientific resources to assess these allocations is currently skewed to such an extent that developing countries are unable to bargain fairly.

The larger question raised by this list is about the rules or principles upon which allocation is to be carried out in each of these four areas. I do not have space in this chapter to go through each of these systematically. What I wish to dwell on here are the economic, quasi-economic, and noneconomic arguments that seem to be involved in all or most of these four allocation issues, and that (apart from their intellectual merit) can be seen as representative of fundamental ways of characterizing not only the allocation issues, but also the worldviews—views of the world—held by the proponents.

Roughly speaking, there are five arguments or positions, which sometimes overlap, and which, as I say, are invoked at different points in Shue's list. Economic arguments have pride of place here, although historically mainstream economists (with one or two exceptions) are relative latecomers to the climate issue. Resource economists and economic geographers had an early (1970s) interest in the impacts of extreme weather or climate events, but there was little interest until more recently in the application of economic tools to any environmental issue, let alone climate.

Utilitarian Market Argument

One economist who did take an early interest in the climate change issue was William Nordhaus, a Yale economist who was involved in considerations of a range of long-term environmental issues early in the 1980s and published influential papers on the economics of greenhouse-gas reductions at the beginning of the 1990s.[21] Versions of these papers also appeared in magazines with widespread and influential readership such as *The Economist*.[22] A somewhat more "risk-averse" version of cost/benefit analysis was proposed by Edward Barbier and David Pearce in their 1990 article "Thinking Economically about Cli-

[21] The framework and discussion in this section owe their genesis and some of their labels to Peter G. Brown, "Fiduciary Responsibility and the Greenhouse Effect," in *Papers from the First International Conference on "Ethics and Environmental Policies,"* ed. Corrado Poli and Peter Timmerman (Padua and Toronto: Fondazione Lanza and the IFIAS Human Dimensions of Global Change Programme, 1993); see also *The Boston Globe*, cited by Peter G. Brown in "Climate Change and the Planetary Trust," *Energy Policy* 20 (1992): 208–22. William Nordhaus, "The Cost of Slowing Climate Change: A Survey," *The Energy Journal* 12, no. 1 (1990): 37–65, and "To Slow or Not to Slow: The Economics of the Greenhouse Effect" (paper presented to the Annual Meetings of the American Association for the Advancement of Science, New Orleans, February 1990).

[22] For example, William D. Nordhaus, "Greenhouse Economics: Count before You Leap," *The Economist*, 7 July 1990, 21–24.

mate Change."[23] It is some indication of the growing influence of economists in the climate debate that in 1991 the British government set up a Centre for Social and Economic Research on the Global Environment (CSERGE), much of it climate-related, and directed by Pearce.

The main ethical criterion of this approach is cost minimization (also describable as efficiency or a refusal to waste scarce resources). We should maximize present utility, measured as ratio of the future benefits of present actions taken to the present costs of those actions. Standard economic modeling techniques are applied to alternative control strategies, which are to be weighed by examining their implications for the consumption (or real income) of different generations.

Nordhaus admits throughout his papers that he is making bold assumptions in the face of widespread uncertainty, some of which include future discounting, unmeasured externalities, a nature with no intrinsic value, and a relatively smooth period of onset of climate change, as well as minor impacts of physical change. A quotation gives the flavor: "Many valuable goods and services escape the net of national income accounting. Among the areas of importance are human health, biological diversity, amenity values of everyday life and leisure, and environmental quality. No one has done the sums here, so it is impossible to say whether the costs of climate change will be large or small. But every time I read of a new deadly tropical virus, I wonder whether humanity could do with a little less biodiversity."[24] Since the measurable economy is what counts, Nordhaus (whose focus is essentially American) argues that agriculture and natural resources are dwindling fractions of the economy and that the impacts of climate warming on the U.S. economy will be minimal.[25]

If emissions reductions are still to proceed, a Nordhausian approach would suggest that the first 5 percent or so of emissions reductions in developed countries should be undertaken immediately, because the costs are relatively low, but that an ambitious project should wait until there is more scientific agreement, or until the signals of global warming are unmistakable. Allocation is primarily assumed to be market-based, and that means allocation by preference and ability to pay.[26]

---

[23] Edward B. Barbier and David W. Pearce, "Thinking Economically about Climate Change," *Energy Policy* 18 (January–February 1990): 11–18.

[24] Nordhaus, "Greenhouse Economics."

[25] For a discussion of agriculture's dwindling role (Nordhaus refers to it as 3 percent of GNP), see Leslie Roberts, "Academy Panel Split on Greenhouse Adaptation," *Science* 253 (1991): 1206.

[26] See Henry Shue's chapter for a discussion and straightforward refutation of this assumption.

Associated with this model is the position that delaying the response to climate change until more information becomes available is the most rational option. There will be more income in the future deriving from more productive kinds of investment, income that can then be spent on more efficiently targeted abatement strategies. This position has received support in part because of the uncertainties associated with the scientific models.[27] Barbier and Pearce, however, within the same position, argue that there is a "waiting cost" for delay—among other things, the future costs for much increased action will be substantially higher.

In a recent modeling exercise, Nordhaus concludes:

A massive effort to slow climate change today would be premature given current understanding of the damages imposed by greenhouse warming. At the same time . . . we must be ever alert to the possibility that the vast geophysical experiment being undertaken by humanity may trigger catastrophic and irreversible changes in droughts, monsoons, ocean circulation, river flows, and other climate-related systems. Economics does not rule out these consequences. If scientific evidence indicates that calamitous consequences are likely to accompany global warming, then our economic models will not only signal that a strenuous effort to slow or prevent future climate change is necessary but help devise the scope and timing of policy responses. Our future lies not in the stars, but in our models.[28]

This orthodox position relies on the belief that economic models and economic activities are both innocent and immediate assessors of appropriate information. Modest insurance is the best policy in the short term, coupled with investment in improved science. The gamble on "too little, too late" is worth taking.

"Realist" or Hobbesian Commons Model

An overlapping approach, which shifts from individual motivation as allocated through the market to national motivation as allocated through power relations, is the realist, Hobbesian, or "zero-sum" game approach, introduced into the debate by Thomas C. Schelling in

[27] A major debate over this issue was conducted in 1991–92 in the pages of *Eos*, a weekly newspaper of the American Geophysical Union. The debate was triggered by a study of how little difference a ten-year delay in cutting emissions would make, by using a simple climate-ocean model. See "When to Act on Climate Change: The Debate Continues as Action Begins," *Global Environmental Change Report* 2 (August 1992): 1–3.

[28] William D. Nordhaus, *Managing the Global Commons: The Economics of Climate Change* (Cambridge: MIT Press, 1994), p. 6.

a National Academy of Sciences Report in 1983 about strategies for dealing with carbon dioxide.[29] Schelling puts the climate issue into a "tragedy of the commons" perspective, but he opts for the neoconservative interpretation of that model, namely, that mutual coercion to prevent self-interested behavior cannot be mutually agreed upon.[30] He argues that because climate change is a commons problem, national states will attempt to maximize their benefits while shifting the costs to others. It could be seen as the political adjunct to the standard economist's model, since all the familiar problems of "rational self-interested behavior" are present.

In June of 1990, Schelling remarked: "Any nation that attempts to mitigate changes in climate through unilateral action pays the costs alone, while sharing the benefits with the rest of the world . . . [therefore] the chances of a global fuel compact are today remote."[31]

Brown notes in his review of Schelling's position that it was ironically at that moment that the nations of the world were sitting down to do precisely that.[32] This would not necessarily invalidate the hypothesis, since the first half of the sentence (the costs of unilateral action) might well explain why there was pressure to get multilateral action. What Schelling's position has a harder time explaining are the actions of the Netherlands and the Scandinavian countries, who have unilaterally engaged in strategies to reduce climate change.

Here we can perhaps invoke again Hampson's concept of "coalitions of the like-minded," or consider the possibility that countries like the Netherlands are gambling that the technological innovations associated with emissions reductions, and so on, will repay their early commitment.[33] There is also intense pressure on governments hosting international meetings (like the Conference of the Parties in Berlin in 1995) to have some positive outcome to the proceedings.

---

[29] For a discussion of realist models, see Steven Forde, "Classical Realism," and Jack Donnelly, "Twentieth Century Realism," both in *Traditions of International Ethics*, ed. Terry Nardin and David R. Mapel (Cambridge: Cambridge University Press, 1992). Thomas C. Schelling, "Climate Change: Implications for Welfare and Policy," *Changing Climate: Report of the Carbon Dioxide Assessment Committee* (Washington, D.C.: National Academy Press, 1983), pp. 449–82.

[30] The original paper by Garrett Hardin, "The Tragedy of the Commons," *Science* 162 (1968): 1243–48, neglects the fact that traditional societies have developed mutual rules to conduct their activities to prevent commons tragedies. Enforcement of these rules can be maintained by strict inclusionary and exclusionary boundaries to avoid "free riders."

[31] *The Boston Globe*, cited by Brown in "Climate Change and the Planetary Trust."

[32] Cited by Brown, "Fiduciary Responsibility."

[33] Hampson, "Climate Change."

Equitable Commons Model

Although as outlined earlier, there have been difficulties in developing a "Southern perspective" on climate change, more sophisticated thinkers and practitioners in the South (especially in India) have adopted a critical stance toward the prevailing orthodoxies of the North.[34] There are a number of variants—one of which I discuss in more detail later—but one of the best known, and most articulated positions is that of Anil Agarwal and Sunita Narain, which could be labeled the "equitable commons" model.[35]

What makes this model similar to Schelling's is that the national state is taken to be the principal actor; what makes it different is that the states of North and South are to act on the basis of rules of equity. The Report of the Commission on Developing Countries and Global Change notes that "international concern for environmental change is producing a normative order that orients investments, commercial movements, and technological relations."[36] It argues that the foundation of this normative order is "the freedom for communities and nations—within a universally accepted framework that prescribes penalties for harming another community or nation—to control the use and management of their natural resources and thereby determine their own form of economic and social development."[37] There is no notion here that national self-interest (like individual self-interest) might be problematic.

The main position espoused by Agarwal and Narain is that global greenhouse-gas emissions should be parceled out equitably on a per capita basis, according to calculations of per capita command over sources and sinks (as owned through the nation state), supplemented by a just and fair share of common oceanic and atmospheric sinks. What they argue is that the current calculations are biased toward the North, and that appropriate calculations can be used as the basis for comparing national contributions to global warming. Because the globe is admitted to be a market (and not a moral) commons,[38] tradable permits within an overall global quota would be permitted.

[34] See "Perspectives from Developing Countries: Conference Statement" and "The Imperative of Equity."

[35] Anil Agarwal and Sunita Narain, *Global Warming in an Unequal World: A Case of Environmental Colonialism* (New Delhi: Centre for Science and Environment, 1991).

[36] Commission on Developing Countries and Global Change, *For Earth's Sake*, p. 68.

[37] Ibid.

[38] A moral commons is usually a closed-access system ordered by the complex interplay of detailed local knowledge with mutually understood ethical restraints. See Fikret Berkes, ed., *Common Property Resources: Ecology and Community-Based Sustainable Development* (London: Belhaven, 1989).

Fiduciary Trust

The fiduciary trust approach allows for economic efficiency, but this is subordinated to a larger goal, along lines analogous to the management of a trust.[39] The argument is that we are trustees of the Earth, and as trustees we have a responsibility in each generation to preserve the Earth's natural and human heritage at a level at least as good as that which we received.

Among the assumptions of this approach that go against the grain of approaches (1) and (2) are that some things are beyond price (sacredness and profaneness can be invoked in this approach). There is also a strange presumption that in some cases the future and the past may be more important than the present. This requires (among other things) the conservation of options, a strong risk-averse strategy especially for potentially irreversible loss, and the application of "precautionary" and "prudential" principles. A notion of "responsibility" is also introduced.

It is not clear in this approach where the boundaries for this invocation of entrusting are, and this is especially important for the demarcation of what is to be kept outside economic considerations. To put this another way: exactly what is being entrusted when we say we are committed to a "planetary trust?" Is it the capacity of the earth to sustain itself, to sustain human life, to sustain human life at least at our current level of flourishing, or what?

Earthrights

The fiduciary trust model overlaps what I call an "Earthrights" model, which turns away from immediate concern with human well-being and attempts to argue on behalf of the flourishing of all beings on the earth, as well as the Earth itself. Various calculations have been made, for example, of the amount of the Earth's productivity from photosynthetic activity now being harvested by human beings for their own benefit—we are now taking approximately 25 percent, which will rise to 40 percent as the population doubles. Unless the burden of human activities drops substantially, the already steep curve of species extinction will only grow steeper.

Do other species and the Earth itself have rights or values apart

[39] The basic document is Edith Brown-Weiss, *In Fairness to Future Generations: International Law, Common Patrimony, and Intergenerational Equity* (New York: Transnational Publishers, 1989); also Edith Brown-Weiss, "The Planetary Trust: Conservation and Intergenerational Equity," *Ecology Law Quarterly* 2, no. 4 (1984): 495–581. Also see Brown, "Climate Change and the Planetary Trust."

from human beings? Various "deep ecologists" and some religious figures argue that they do.[40] If that is so, then allocation questions are not
merely concerned with the human-to-human. Perhaps to advance on
the "fiduciary trust" model, the boundaries of what can be economized should (according to an "Earthrights approach") also be supplemented by the drawing of the boundaries of what should not be
humanized.

CASE STUDY: PERSON AS PACKAGE

To look at the interplay of some of these five positions in more detail,
I have chosen to focus on a particular theme that is quite prominent in
the literature—what I call the "person as package" model.

In 1992, Dr. Jyoti Parikh of the Indira Gandhi Institute of Development Research, India, wrote a critique of the current IPCC reports,
blending both scientific and ethical claims together, and drawing upon
a very long and complicated debate in the energy and climate impacts
literature.[41] As only one example of the blend of science and ethics, she
brings up the issue of when one starts counting carbon dioxide emissions. It is generally in the interest of northern (that is, OECD) countries to start as recently as possible, with current percentages, which
would mean that OECD countries are responsible for 70 percent of
current emissions. Developing countries—borrowing from work done
by Florentine Krause and others[42]—argue instead that, since carbon
dioxide has accumulated in the atmosphere over time, we should be
calculating the available carbon dioxide budget since the beginning of
the Industrial Revolution, in which case industrialized countries are
responsible for 85 percent or more of emissions, and, in fact, have
used up most of their quota for the foreseeable future.

Parikh also notes that the so-called "stabilization" plan favored by
the IPCC, which is designed to keep emissions at 1985 levels (5.15
billion tons of carbon), calls for a 59 percent global reduction in emissions by the year 2025, but this is calculated by taking the numbers

[40] Two different perspectives—from a Christian viewpoint, John B. Cobb Jr., *Sustainability: Economics, Ecology, and Justice* (Maryknoll, N.Y.: Orbis Books, 1992);
from a Buddhist viewpoint, Gary Snyder, *The Practice of the Wild* (San Francisco:
North Point Press, 1990). A great deal of recent theological work specifically related to
climate change has been carried out by the World Council of Churches, for example,
Accelerated Climate Change: Sign of Peril, Test of Faith (Geneva: World Council of
Churches Study Paper, 1994). Also James A. Nash, "Ethical Concerns for the Global
Warming Debate," *The Christian Century*, 26 August–2 September 1992, 773–76.
[41] Jyoti K. Parikh, "IPCC Strategies Unfair to the South," *Nature* 360 (1992): 507–8.
[42] Florentine Krause, Wilfrid Bach, and Jon Koomey, *Energy Policy in the Greenhouse*
(El Cerrito, Calif.: International Project for Sustainable Energy Paths [IPSEP], 1989).

from a "do-nothing" scenario for 2025. In other words, the base-line
for reduction is worked backward from a worst-case scenario for a
continuing and indeed widening inequitable distribution of energy use.
This gives developed countries an even bigger advantage, because their
allocation of future shares of the global "carbon pie" would come
from an even bigger baseline percentage of emissions than now. Parikh
argues: "The IPCC document shows great concern (to the point of
paranoia) about potential growth in emissions from the South, over-
looking the fact that total emissions from the North, despite its
smaller share of the global population, are higher than those of the
South. The South needs time to develop, and the North must accept
that, for a certain time, the burden of the stabilization scheme must be
borne by all those and only those who emit more than the world average emissions in 1985, say 1.1 tonnes per person."[43] (Parenthetically it
should be noted that currently in North America emissions are over 5
tons per person, as compared to approximately 0.29 in Africa or 0.19
in South and East Asia.)

The last sentence of Parikh's is revealing, because it is a slight varia-
tion of the Agarwal position (the Equitable Commons Model), since it
focuses only on per capita emissions, without regard for each person's
share of sinks. Everyone should be allocated an equal package of permissible emissions. This is also in contrast to the three most popular
(among developed countries) alternatives: allocations of entitlements
in proportion to present emission levels, in proportion to expected
future emissions levels, or according to GNP levels.

There are a number of rhetorical variants of this South perspective
that also focus on the person as package (or "bearer of emissions
rights"). An interesting locus is Paul Kennedy's *Preparing for the
Twenty-first Century* (1993), where global demographics suddenly
and somewhat gratifyingly appears as a serious policy issue.[44] In its
review of the book, *The Economist* makes the following criticism:
"We are also told that the Average American baby represents 280
times the environmental damage of a Chadian or Haitian baby, because its level of consumption throughout its life will be so much
greater. This is an extraordinary misinterpretation of the fact that the
American baby's mom and pop have hitherto been more productive."[45]

This is, of course, the counterargument to the position I described a
moment ago emanating from developing countries (the internal refer-

[43] Parikh, "IPCC Strategies," p. 508.
[44] Paul Kennedy, *Preparing for the Twenty-first Century* (New York: HarperCollins,
1993).
[45] "Millions of Mouths to Starve," *The Economist*, 3 April 1993, 84.

ence in Kennedy's book is to the Ehrlichs and their 1990 book, *The Population Explosion*).[46] The counterargument goes back at least to Julian Simon's *The Ultimate Resource* (1981), which is cited by Kennedy.[47]

I have personally heard more than once (usually after the official day was done) the following statement, variously phrased: "If it weren't for Northern industrial development, no one would have ever found out about global warming until it was too late." The witty paradox, of course, is that of the problem generating its own solution; but duller variants have raised the questions of (1) whether productive developments in the First World which assisted the Third World are to be discounted (for example, the knowledge base from which all people now operate), and/or (2) how are we to determine whether some emissions are more equal than others?

We could argue that the previous emissions were investments in today's well-being (for some), rather than being purely burdens. They were then more productive than emissions that were not investments. Productive of what, one might ask? It could be counterargued (by developing-country defenders) that what Western consumption is productive of is more consumption. Another counterargument, taken by the Global Commons Institute, is to turn the tables by assessing the efficiency of use of carbon dioxide relative to GDP. In this analysis developed countries come off very poorly due to their waste of energy.[48]

The original idea of weighing and comparing people according to ratios as a tool in ethical arguments, particularly in the population debate, is well known, but I believe it to be dangerous. There has already been substantial controversy over remarks made in a CSERGE paper about comparing the value of a life in the nondeveloped world to a life in the developed world through willingness-to-pay criteria.[49]

It is presumably possible to come up with some sophisticated equation representing the package, such as Paul Ehrlich and Anne Ehrlich's impact equation $I = PAT$ (Total Impact equals Population times Affluence times Technology).[50] Or one that would come up with a per cap-

[46] Paul R. Ehrlich and Anne H. Ehrlich, *The Population Explosion* (New York: Simon & Schuster, Touchstone, 1990).

[47] Julian L. Simon, *The Ultimate Resource* (Princeton: Princeton University Press, 1981).

[48] See among other papers, "Climate Change Economy and The Global Commons: Global Considerations of Efficiency, Equity, and Ecology," Global Commons Institute, London, 1994.

[49] See Samuel Fankhauser, "Global Warming Damage Costs: Some Monetary Estimates," Working Paper GEC 92–29, CSERGE, London, 1992.

[50] Ehrlich and Ehrlich, *The Population Explosion*.

ita package relating together (1) lifetime consumption; (2) lifetime production; and (3) lifetime environmental damage from all sources including baseline functions such as breathing, eating, excreting, methane production, recycling of bodily parts, and composting.

The physical part of the package is almost certainly the easiest to work out, and there has been lots of work on basic caloric needs and so on. But the production and consumption parts of the package immediately run up again against all the interpretative issues. Ecologically, for example, we can ask whether one of the reasons for the long-term success of northern countries has been the resilience of temperate ecosystems, which is not the case for much of the Third World. Is the average northerner more productive because of innate wisdom or getting there first, or as a result of a history of sucking resources out of the Third World? For example, current consumption patterns in the West are crucially dependent on extraordinarily—some would say punishingly—low commodity prices. The more general point is that once one starts into the "person as package" argument, it is hard to know where to stop, what should be made part of the package, and so on.

More troubling for developing country defenders, the *per capita* emissions model, unless it is tied to some kind of mutually satisfactory production function, opens up the prospect of reducing people to emitters of gases, and this kind of reductionism plays into the hands of those, for example, who wish to reduce the population debate to a question of numbers. This seems to me to be, in a different sense, at least dangerous if not unethical.

In their variant, what Agarwal and his colleagues have finessed is the question of why "person as package" rights, extended to include not just emissions but control over sinks, should be given over to the state to bargain with. This elides basic questions such as how much control ordinary people do have over forests, mangrove swamps, and so on, in developing countries (or in developed countries for that matter). Is a peasant in Latin America really benefiting from large-scale deforestation for cattle ranching? If individual per capita emissions are what is important, why can I not trade them individually with someone else? Why must I go through the state to do so?

In fact, what we currently find is that "joint-implementation" strategies are not only being negotiated by national states, but by borderless transnational corporations and other institutions that have "moved to the global" in such a way that they are organizing present and future market share.

Again, even if we define people as sources of gas, and as quasi-proprietors of some form of sink as designated by property rights or the state, why should we stop there? Why should we not invoke the usefulness of emissions as *The Economist* does, and then the entrepreneurial skills and the command over societal resources of the individual could be counted?

To conclude this discussion on a practical note, it appears that the move toward some form of per capita emissions targets seems to be gaining ground. In order to ensure that there is no "population bias," there have been suggestions that if a protocol is put into place it should begin at the national state level, and then gradually move toward per capita targets.[51]

ALTHOUGH there are problems with the "person as package" model, it does raise acutely the question of what emissions are used for, and by doing that it crosses what seems to be the most important unspoken barrier in the whole discussion, a barrier that is kept in place by the use of standard economic reasoning about the allocation of resources. Shue, for example, expresses the opinion that some kind of distinction between subsistence (or necessary or "needs-based") emissions and luxury (or "wants-based") emissions is required. As soon as this kind of analysis is allowed, then the barrier to making moral judgments against the North (and to a lesser degree the South?) in detail is broached. The "homogenizing" of economic goods by the market is designed expressly to forbid any other comparisons but those of price, which is supposed to signal these moral issues—if they matter enough people will pay for them. The market is the arbiter "at one remove from reality" of what matters, economically as well as physically.

One small crack in this barrier is provided in the climate change issue by the scientific framework for discussion, which is inexorably moving toward quite specific analyses of what are called "sources" and "sinks." There is a shadowy hope that these would be able to be put into an ecologically "homogenized" frame (for example, by finding a single measure such as greenhouse warming potential [GWP]), in order to complement an economically "homogenized" frame of costs and benefits, and thus result in something like a tradable GWP regime.

For a variety of reasons, this is an unlikely prospect. Scientifically, sources and sinks differentiate in all sorts of characteristics, including their productive and absorptive capacities. Agarwal and Narain have already made much of this differentiation in their arguments in favor

---

[51] For a balanced and influential discussion of the allocation problem, see Michael Grubb, *The Greenhouse Effect: Negotiating Targets* (London: Royal Institute for International Affairs, 1989).

of "nations as packages." There is also a false analogy between economic and ecological systems that has been endemic (at least in ecology) since the 1960s.

As the issues of what constitutes acceptable emissions reduction strategies, what inventory methodologies are to be pursued, and what the indirect costs and benefits of emissions reduction strategies are all become more focused, it seems likely that the pressure to devise an economistic framework for decision making will increase, as will the number of ethical claims and counterclaims.[52] I suspect that the North will have to accept the transgressing of the boundary of economism, just as the South will have to accept the transgressing of the boundary of the state as sovereign arbiter. In the drafting process for the IPCC Working Group III Report on cross-cutting issues—including economics and equity—the tensions between the economists and the ethicists have become overt.[53]

Paradoxically enough, recent global negotiations have been in part successful because of the uncertainties involved. It has been possible for developed countries to ignore quietly or accept tacitly the various claims put forward by developing countries, largely by themselves "going global." Still, as the potential costs and benefits of climate change become clearer, and as the potential winners and losers from global warming become more easily identified, it is likely that the conventions will come under severe stress from further polarizations.

Nevertheless, it is important to insert at this point a caveat from the scientists. Over the past ten years, thanks to chaos and complexity theory, there has been a revolution in thinking about complex natural systems, including both the natural ecology and the climate. It is now fairly widely recognized that we are barred in significant ways from predicting in any kind of detail the future of these systems.[54] For example, although the results are controversial, recent deep ice-core sam-

---

[52] For example, Chichilnisky, Heal, and Starret argued at the IPCC Working Group III Nairobi Workshop in 1994 that the initial international distribution of entitlements to produce carbon dioxide will affect the efficiency of a tradable permits system, and that equity issues are therefore not separable from the market. This has to do with the complexities of a public good (atmospheric carbon dioxide) being consumed by all, but produced by a finite number of agents (if the national state is bundled together as the agent). See Graciela Chichilnisky, Geoffrey Heal, and David Starret, "International Emissions Permits: Equity and Efficiency" (November 1993, tabled at Nairobi Workshop on Equity and Social Considerations, 19 July 1994).

[53] The author and others are investigating this issue in more detail in a current research project at Concordia University titled "Ethics of the Embedded Market."

[54] See C. J. Edwards and Henry A. Regier, eds., An Ecosystem Approach to the Integrity of the Great Lakes in Turbulent Times (Ann Arbor, Mich.: Great Lakes Fishery Commission, Special Publication 90–4, 1990) for an example of the practical implications of this new approach.

plings have indicated that the Earth's climate system, when under significant stress from warming, can fluctuate quite dramatically.[55] Virtually none of the models and discussions concerning global change give much weight to the possibility of "surprises" or "non-linearities"—partly because people are afraid to present themselves as apocalyptics, and also because nonlinearities tend to "sweep the board" free of plausible, comfortable discussion.

Elsewhere I have argued that this prospect substantially alters (or should alter) the managerial assumptions upon which modern society has built its legitimacy: assumptions of prediction, control, and expert judgement.[56] Ludwig, Hilbourn, and Walters made a similar review of the failure of the concept of "maximum sustainable yield" and concluded that "we shall never attain consensus concerning the systems that are being exploited."[57] This resignation to uncertainty helps to explain a growing movement toward the preparation of declarations of principle or "charters." I suspect that in the face of future uncertainty, citizens, institutions, and governments are having to reconsider what will constitute "appropriate practice." If detailed management plans are of less and less use, how is trust in governance to be sustained, especially if ecosystems are going to be increasingly prone to swings of unpredictable behavior?

In this light, the arrival of "prudence" and "trust" as topics of contemporary interest in political science is no accident.[58] The "precautionary principle" is now well enshrined in the enviro-political lexicon, and a global precautionary principle is also likely to emerge over time, perhaps along the lines of Hans Jonas's "ontological imperative"— that is, "Do not compromise the conditions for an indefinite continuation of humanity on earth."[59] Jonas notes in this context: "The crucial point in all this is that the nature of human action has changed, and with it the focus of ethical theory . . . we must see that responsibility with a never known burden and range has moved into the center of political morality."[60]

The largest ethical issue of all that we face remains the movement

[55] Greenland Ice-Core Project (GRIP) Members, "Climate Instability."

[56] Peter Timmerman, "Emergent Measures: Ecosystem Management and Public Participation under Uncertainty," *Alternatives* (forthcoming).

[57] Donald Ludwig, Ray Hilbourn, and Carl Walters, "Uncertainty, Resource Exploitation, and Conservation: Lessons from History," *Science* 260 (1993): 17–36.

[58] John Dunn, *Interpreting Political Responsibility* (Princeton: Princeton University Press, 1990).

[59] Hans Jonas, *The Imperative of Responsibility* (Chicago: Chicago University Press, 1984).

[60] Ibid., p. 122.

toward a utopian conception of "managing the planet" by, for, and on behalf of human beings.[61] Disguised behind the standard approach of "the future as usual" is the following kind of disquieting fantasy (from a paper by William Nordhaus):

> A promising new approach to the threat of greenhouse warming is to use our brains to find a way to offset greenhouse warming through climatic engineering; this is the global equivalent of turning on an air conditioner. . . . Careful analysis of these proposals is just beginning, but a number have already been identified that appear much more cost-effective than plugging the oil wells and shutting down the coal mines. One approach would be to create a sunscreen by sending tiny particulates into the stratosphere to cool the earth. These particles could be shot up with 16" naval rifles, lifted by hydrogen balloons, or deposited by tuning the engines of aircraft to burn somewhat richer than normal.[62]

Homeopathic pollution remedies are evidence of a kind of engineered solution to our problems that seemingly moves the issue from the ethical to the technical sphere, but that disguises the larger ethical question of whether technological optimism of this kind is the advance guard for a scheme of planetary management, which will be seen to be necessary if things go well, and even more necessary if things go badly. Emergencies have historically been moments when not only are existing structures called into question, but insecurities also beget new encroachments of the powerful in the name of necessity.

This move toward the question of how to "manage" the planet—the last great global power grab—raises in acute form the alternative and equally pressing question—"Can we manage to make it?"[63] Global change, and the maneuvering around it, are the place where the answers are likely to be worked out first.

[61] "Managing the Planet" was the title of the September 1989 issue of *Scientific American*.

[62] William Nordhaus, "The Cost of Slowing Climate Change," p. 61. Most telling is the footnote to this passage in the original: "The technological options discussed here were derived from communications with Robert Frosch of General Motors, to whom I am most grateful for clarifying several issues."

[63] I owe this formulation to Steve Rayner.

Conclusion:
Liberalism Is Not Enough

JUDITH REPPY and FEN OSLER HAMPSON

THE environmental changes that have already occurred as a result of human activities, and the much greater changes that loom on the horizon, challenge our collective ability to alter our behavior. Adaptation to the changing environment will surely take place, and human beings may even develop measures to prevent or mitigate environmental change. The argument of this book has been that any such response to global environmental change must incorporate a concern for social justice and a respect for the well-being of the ecosystem. Ethical issues are not "add-ons" to the policy issues raised by environmental change, they are fundamental to the framing of the problems to be addressed and the search for acceptable solutions. Absent these moral concerns, we run the risk of devising environmental policies that will perpetuate inequity within and among societies and further damage the relationship between human society and the larger ecosystem.

Several main themes emerge from the chapters in this book. First, a focus on global environmental change and long-term sustainability requires rethinking our theories of justice. Second, our analysis should allow moral standing for communities as well as individuals. There is a serious question whether we should extend the same arguments to other sentient beings or to the ecosystem as a whole. Third, there is a similar difference of opinion about the potential for reforming the practices of the modern state to provide adequate protection for the environment and full participation by all peoples. Whereas some argue that fundamental changes in the nature of the state and the international state system are essential, others believe that the transition to greater social justice and respect for the environment should not be held hostage to such sweeping changes—hence the need for transitional ethics. And fourth, the production and use of scientific knowledge should be democratized to allow participation by representatives of less developed countries and sensitivity to the interests of marginalized communities.

The authors represented in this book have different views regarding these ideas, but collectively their contributions provide a searching analysis of the issues and point the way to further questions.

What Do We Require of a Theory of Justice?

In discussing the limitations of traditional concepts of distributive justice, Henry Shue, Will Kymlicka, and Wendy Donner underscore our need to think about traditional social boundaries, such as those of the national state, in a more flexible way because nature is indifferent to them. Ecoethics must deal with social, cultural, and ecological systems taken together; in particular, people should be aware that when they are making environmental decisions they are also making explicit ethical choices. The conflicts that underlie these choices are not necessarily between different kinds of moral reasoning, but at a more fundamental level, between different conceptions of what constitutes the moral community.

Arguing that global environmental change requires a new ethical approach based on fault-based standards of justice, Shue takes the view that responsibility for correcting injustice should generally track causal responsibility. In the case of global warming, for example, he suggests that responsibility for addressing the problem should be assumed by those who are in fact responsible for creating it. At the same time, Shue indicates that we should recognize that some countries may be too poor to contribute to the solution and, therefore, richer countries will have to assume a correspondingly greater share of the burden.

Much of the international policy debate about global warming has centered on the relative costs of mitigation versus adaptation strategies, with some economists arguing that adaptation may in fact represent the most efficient response to the uncertainties of global warming. The most efficient strategies for dealing with global warming, however, may not be the most just. For example, Bangladeshis who are threatened by rising sea levels will be the least able to pay for adaptation because their country is one of the world's poorest. Shue argues

that greater attention should be paid to mitigation and alleviating the
stresses of those who are most likely to experience the greatest hard-
ship from global environmental change. The principle that rich countries have an obligation to help poor countries is reflected in the Stockholm Declaration of 1972, the Montreal Protocol for the Protection of
the Ozone Layer, and the Rio Declaration on Environment and Development, which makes a case for "common but differentiated responsibilities." Shue argues, however, that the ethical standards reflected in
these international declarations and agreements are still no-fault, that
is, they refer to an ability to pay but do not assign blame for the
problem. This is somewhat surprising because fault-based, or "polluter pays," standards are enshrined in the domestic laws of many countries as well as within the OECD and the European Union. Shue notes
that efforts to establish fault-based standards are virtually guaranteed
to become embroiled in controversies about causes. There is also the
difficulty of assigning historical responsibility for a problem: after all,
at what point in time does the assignment of blame and responsibility
stop? Yet, he notes, "It would seem strange always to act as if the
world began yesterday, by never attempting to assess fault."

Environmental change raises issues of justice within as well as
among states. Will Kymlicka argues that theories of distributive justice
between states do not guarantee the justice for substate communities,
particularly in the case of indigenous peoples. In a world in which
there are stark disparities of wealth we cannot avoid questions of how
to go about fairly allocating people and land. Even the new philosophy of ecocentrism has important distributive implications; if certain
areas are to be preserved from development, then it is important to
ensure that the costs of doing so do not fall disproportionately on the
poor. Still, neither mainstream conceptions of social justice nor more
recent environmental theories have tackled the dilemma of how to reconcile competing communal interests when the issue is how and
whether to exploit natural resources in ecologically fragile zones.

Kymlicka takes up the challenge of how to incorporate the concept
of community into the concept of social justice. He notes that there
are often disparities in wealth and power at the local level and that
indigenous peoples usually end up with the short end of the stick.
How should these inequities be addressed? According to Kymlicka,
indigenous peoples represent minority cultures and are therefore entitled to the special rights and resources needed to preserve their cultures. But their claims also have real limits. Although indigenous peoples are entitled to self-government, this right does not absolve them
from their own obligations to the wider community and to the envi-

ronment itself. Kymlicka concludes that although we need to develop
an approach to justice that is sensitive to community, it should not
elevate the moral standing of any community into an absolute. The
interdependent nature of environmental problems and the fact that
they do not respect borders means that everyone has an interest in
determining how resources are used and whether they are used wisely.

Wendy Donner provides a review of those theories of justice that
extend moral standing to other species or to the ecosystem as a whole.
Animal rights theories suggest that all creatures have intrinsic moral
standing or worth. This creates a practical problem because, as Donner explains, "as more and more individuals are granted standing, it
becomes more and more difficult to reach any practical solution to
moral dilemmas."

Holism or ecocentrism makes the environment the prime bearer of
inherent value; the biotic whole is therefore more important than the
interests or fate of individual species, including the human species. In
fact, when our value as a species is assessed in terms of our impact on
the environment, we fall at the low end of the value scale. Deep ecology has no base of political or economic theory; thus it attributes
human poverty and overpopulation in the Third World to moral irresponsibility rather than to causes rooted in structural underdevelopment. These theories are generally unsatisfactory because they fail to
offer guidelines on how to balance competing claims among very different kinds of moral agents, and in their devaluation of human life,
they lead to conclusions that are sharply at odds with our usual moral
intuition. Environmental ethics has to recognize that human beings are
moral agents who need guidelines for reconciling conflicts between human and environmental values.

Ecological feminism with its emphasis on such values as care,
friendship, and reciprocity in personal relationships provides a useful
alternative to the androcentric values of current society. In ecofeminism all of nature has value, and therefore human beings should be
careful about imposing their own value hierarchies upon it. Yet ecofeminism, like the other theories Donner discusses, provides little guidance for conflicts where claims are equally valued or for practical action with regard to the environment.

Iain Wallace and David Knight observe that new concepts of space
and community are informing discussions of justice and environmental
change. Against claims of universal norms they argue that the geographically bounded community has a moral particularity of its own,
one rooted in its individual history and place. In making the case that
just solutions to resource management require a balanced and equal

relationship between core and periphery, Wallace and Knight echo many of Kymlicka's concerns. But they go one step further, suggesting that public policies toward peoples whose livelihoods are at stake in environmentally stressed regions should not bear disproportionate losses because their industries fail to meet new environmental standards, standards often promulgated from the privileged core. If we value "place," then we should also value those who live within it.

We can conclude that the demands of social justice are inseparable from our responses to environmental change and our respect for the ecosytem: attention to one necessarily implicates the other. There is a general consensus among our authors that traditional liberal theory is an imperfect framework for evaluating competing moral claims that arise in the context of environmental change. The standard liberal argument, which assigns primacy to individual human rights and preferences, has no answer to the questions that arise when whole communities are threatened by environmental change or by the policies proposed in response to such change. To the extent that human societies are grounded in an essential way in the physical and biological places they inhabit, there is no satisfactory way of compensating them for the destruction of those places or the elimination of their traditional means of livelihood. Similarly, a concern for social justice for individuals does not suffice when the cultural identity of the group is threatened. This is not to argue that the interests of the community or group should automatically trump the moral claims of the individual, but simply to call for a theory of justice that allows communal values and future generations to be considered alongside the rights of the individuals living today.

Further questions remain, however. Are there general principles or metarules of moral reasoning that we can invoke to decide between competing interests when all have a valid moral claim? Are there any claims that have absolute priority? And perhaps most important, how can we provide a voice for the poor, for marginalized communities, for future generations in the decisions that affect the environment?

WHAT ROLE FOR THE STATE?

Global environmental change simultaneously ignores state boundaries and challenges state capacity to act, individually and in concert with other states. In a world made up of national states, the character of the state and the current structure of international society cannot help but affect the possibilities of achieving a new global environmental ethic, one that is equally sensitive to the needs of community and to the other species who inhabit the planet. Although the authors in this book differ in their views of the state, all are critical about the capacity of the modern state to respond adequately to the moral challenges of global environmental change.

Is it realistic to speak about social justice and redistributive policies in a world where nations are nominally sovereign and power is unequally distributed? Christian Reus-Smit argues that distributive theories of justice are not adequate for analyzing issues of background justice, which is concerned with the (nondistributable) nature of international norms, rules, and procedures. According to Reus-Smit, most theories of international environmental cooperation are built on an individualist ontology. He argues for a "societal ontology" that recognizes social interests, culture, and intersubjective human practices.

In Reus-Smit's view, the moral purpose of the modern state, which is defined largely in economic or developmental terms, has legitimized certain kinds of activities and proscriptions while delegitimizing others, and thereby structured international environmental negotiations in a fundamentally unjust manner. At one end of the political hierarchy are the producers of scientific knowledge and the privileged consumers of that knowledge, namely, the industrialized nations. At the other end are the indigenous peoples, who have been mostly unsuccessful in influencing the terms of the debate and the basic norms and principles behind recent international environmental agreements. He therefore believes there is need to redefine the moral purpose of the modern state if we are to develop international institutions that will be more inclusive of underrepresented points of view. Historical conditions may favor doing this now, but the impetus will have to come from below.

Reus-Smit's critical view of the modern state as an obstacle to social justice is also discussed by Joseph Camilleri. Camilleri argues that the nation-state has had both positive and negative impacts on global environmental change. For him the state is an arena for social conflict, not the main obstacle to change, as suggested by Reus-Smit. The state faces multiple social pressures and may therefore lack the regulatory capacity to deal with many environmental problems, if competing interests are deadlocked. Camilleri adopts a world-system approach to argue that the fragility of the environment in the Third World is linked to the marginalization of Third World economies and the harsh consequences of structural adjustment policies imposed by Western donor and financial institutions. To be sure, the "core" is not monolithic; there are divisions among elites at both the core and the periphery. But the tensions between "core" and "periphery" affect the way environmental and distributional issues are being addressed internationally and domestically. Camilleri sees normative, institutional, and cultural contradictions in current efforts to address global environmental change. Ad hoc change, he says, is occurring, but as a result of these contradictions, it lacks a normative consensus.

According to Camilleri, there are five requirements for change: (1) achievement of new normative consensus; (2) integration of equity and ecology into economic decision making; (3) a redefinition of the social contract; (4) restructuring of bargains, intra- and inter-state, so they are just; and (5) a revised relationship between states, markets, and civil society. No single approach to institutional change can be successful by itself. Only the emergence of different institutional structures at the micro and macro level can ensure a just approach to environmental decision making.

In response to the question of how a new normative consensus is to be achieved, Smitu Kothari suggests that the necessary changes will only come about through the formation of social movements from below. Like Reus-Smit, Kothari is sharply critical of the role of the modern capitalist state, seeing it as a source of both human injustice and environmental degradation. Patterns of unequal development and social stratification result in natural-resource extraction for the benefit of the few. In many countries, however, social movements are challenging the notion of development as an unquestioned good and seriously pursuing alternative systems of governance that are locally based and informed by a deep understanding of local ecology. The movements seek to preserve cultural diversity along with biodiversity; it is not a question of defending pristine nature, but of "social" nature. To do this requires a shift of political praxis from the state level to communities, along with a simultaneous democratization of the state to open up greater political and social space for movement activity.

There is a consensus in these chapters that if a genuine moral discourse is to occur, new participatory structures are needed to give voice to hitherto unrepresented or underrepresented groups in society. At the same time, the authors stress that greater political participation will not necessarily enfranchise the disenfranchised, particularly if the more powerful interests in global society set the terms. Nor is it axiomatic that greater participation will lead to policies that respect the environment. Paradoxically, the state is seen both as an instrument of oppression and as the most plausible protector of national resources and the environment against the global forces of markets and industrialization. The challenge facing the critics of the developmental state is to chart the transition to a more just society in greater detail.

SCIENCE AND SOCIAL JUSTICE

International debate and discussion about global environmental
change have been dominated by two communities: scientists and environmental nongovernmental organizations (ENGOs). Social scientists
have paid growing attention to the political influence exercised by
these two groups of actors in international relations. At one level, scientific discourse translates into an ethically neutral discourse because,
by its claims to universality and objectivity, it denies the existence or
importance of the national state and other identities or group affiliations. At another level, science is crucial to the state because it has
become the adopted discourse of the most powerful and wealthy nations on the earth and the basis and legitimacy for key policy decisions. The final chapters in this book address two interrelated questions: (1) What is the role of science in framing the moral choices of
global environmental change? And, (2) What are the potential dangers
of scientification of moral discourse on global environmental matters?
Sheila Jasanoff takes up the first question in her chapter, "Science and
Norms in Global Environmental Regimes." She first of all points out
that in response to the growing importance of issues like hunger, disease, and global environmental degradation, the world is turning to
science and technology for solutions. As the only kind of "universal
discourse available to a multiply fragmented world," science's appeal
is understandable. But we should be skeptical, says Jasanoff, about
science's claims to universality. This is because science has owners,
among them governments and corporate interests, and these are unlikely to challenge the political and normative structures in which they
are embedded. In assessing the claims of experts, asserts Jasanoff, we
must not only ask how issues are framed, but "Who is doing the framing?"

This same question is also addressed in Steven Yearley's chapter on
the role of ENGOs. Yearley begins by noting a growing converging
trend toward the adoption of scientific arguments as opposed to morally based appeals among ENGOs. Although the conservation movement was initially made up of nonscientists, in recent years it has become more precisely scientifically trained. The ENGO role has also
changed to one where ENGOs are increasingly proposing and defending specific policy alternatives. This has led to two contrasting attitudes to scientific authority. Some groups have embraced science as a
way of enhancing their claims to legitimacy on the grounds that scientific claims are morally disinterested and universal. Others have taken
an approach that relies less on science and more on moral appeals to

the inherent "rightness" of their position. On the one hand, resorting to scientific claims as a campaign tactic cannot replace other kinds of considerations—in the case of whales, for example, the importance of moral concerns. On the other hand, appeals that ignore scientific evidence run the risk of undermining broader moral appeals and fostering public skepticism about the environmental movement as a whole.

The movement toward environmental campaigns at a global level has also encouraged some ENGOs to think that their appeals are universal and will be accepted carte blanche by developing countries. This is not so: ENGOs have been viewed with mistrust in many developing countries because their activities threaten sovereignty and a peoples' right to control their own resources. Although sovereignty is not absolute, neither are the claims of some environmentalists. By mobilizing and working with constituencies within the state, ENGOs can legitimize their claims. Indigenous voices are usually more effective than outside ones—and their appeals tend to have greater moral urgency.

But, as noted above, the international negotiations have focused on technical issues. Peter Timmerman suggests that part of the problem with the current debate is that most developing countries are disadvantaged in participating in the scientific debate and lack adequate resources to inventory, monitor, and police agreements that are signed. If we are to move toward the kind of transition ethics suggested by Henry Shue, argues Timmerman, developing countries will have to become full and equal partners in the debate by obtaining better access to science and by developing their own indigenous scientific capabilities. At the same time, appeals for international cooperation—even appeals coming from groups with full representation of scientists from developing countries—cannot ignore the stark disparities in wealth and income in the world and the distributive consequences of global environmental change.

All of our contributors underscore the domination of the current discourse on global environmental change by Western science and by the advanced industrialized countries. There is an urgent need to bring values into the debate and to ensure that the voices from the periphery are not just heard but are involved in decision making.

LEVELING THE PLAYING FIELD

The current politics of justice of the environment is one marked by disparities in wealth, income, access, information, and levels of political participation across and within societies. All our contributors suggest that there is a need to address these disparities by developing new standards of social justice that recognize moral responsibility, confer new rights and legitimacy to the communal and cultural interests of indigenous peoples, and develop new participatory structures of national and international governance. There is also an ethical imperative to bring the rights and interests of the human species into a better defined balance with those of other species who inhabit the planet. There is no guarantee that new concepts of political participation and moral value will lead to a solution to the world's environmental problems. But there is a need for better criteria for steering this process so that the human race does not continue to perpetuate those practices and behaviors which led to the problem in the first place.

Although it is not plausible that we can completely eliminate inequity, there is a moral imperative to try to reduce it. Participatory solutions to global environmental change require real resource transfers of finance, technology, and knowledge to the world's poorest societies so that they can participate more effectively and as equal partners in the development of global solutions to these problems. Without a process that gives moral standing and voice to all, including communal groups who are affected by environmental change, we risk "solutions" that place the burden of adjustment on those least able to bear them. We also run the risk of failure, because developing countries are in a position, perhaps for the first time, to undermine preventive or adaptive measures taken by industrialized countries. If China, for example, continues to industrialize with the speed it has by using relatively inexpensive coal and fossil fuels to meet its energy needs, whatever the industrialized world does to reduce greenhouse-gas emissions will be of little consequence in reducing the likelihood of global warming. Self-interest, in this case, may well be the handmaiden of a morally sound policy.

The proposals and arguments presented here are only a first step. Yet if taken seriously, they will require drastic revisions in how we configure international negotiations on environmental change and the range of policies considered. By invoking norms that value fairness and the protection of biological diversity and sustainability, developing countries can win support in the negotiation for policies that address their concerns for equity and development, for the survival of communities, and for long-term sustainability. This support should come through an appeal to the value systems of elites engaged in negotiations and through pressure by mobilized public opinion and nongovernmental organizations, as Peter Timmerman suggests in Chapter 10.

Such recommendations are not a panacea: new conflicts over policy will arise with the arrival of new voices and new demands at the bargaining table. There is considerable potential for clashes between the value placed on particular interests, over considerations of equity, and between respect for customary practices and a concern for women's rights. But beggar-thy-neighbor policies at the communal and international levels are morally deficient in confronting the new challenges posed by global environmental change. Sharing responsibility, on the other hand, represents the morally preferable course, especially if responsibility is based on a formula that takes into account each society's resources and ability to handle the costs of adaptive and preventive measures. As this discussion has shown, there is no abstract principle of justice that allows us to reconcile the competing claims and values of different communities. Human and planetary survival may well depend upon the development of a new moral sense of obligation that not only goes beyond the traditional boundaries of the nation-state but also respects the communitarian rights of indigenous peoples and other minorities.

Index

Abbey, Edward, 59–60
Action Plan: Sustainable Netherlands (Friends of the Earth), 218
Adas, Michael, 112
Affirmative action, 46
Agarwal, Anil, 23, 235, 238, 240–42
Aiken, William, 59
Allen, Robert, 210
Animal rights/animal welfare, 53–57, 61, 248
Anthropocentrism, 52, 72
Ashley, Richard, 188n, 189–90, 194
Authority, 8; scientific, 114, 164–65, 195–97, 206–9, 219

Bangladesh, Chittagong Hill Tracts, 3, 31–32, 48
Banuri, Tariq, 165
Barbier, Edward, 231–33
Bargaining. *See* Decision making; Standards
Basic Rights (Shue), 102
Beauvoir, Simone de, 67
Beck, Ulrich, 184, 199–200, 219
Beckerman, Wilfrid, 26
Beitz, Charles, 97
Benhabib, Seyla, 83
Bentham, Jeremy, 54
Biehl, Janet, 70
Biodiversity, 75–76, 159–60, 166
Bioregions, 41–42

Biospheric community, 58–59, 92–94
Bohlen, Jim, 202–3
Bolin, Bert, 27
Bramble, Barbara, 205
Brazil: Amazonia, 32, 38–40, 42, 44, 114–15, 142; community in, 133; economic exploitation of, 131
British Columbia, Clayoquot Sound, 3, 53, 93–94
Bruntland Commission Report (*Our Common Future*), 89

Callicott, J. Baird, 53, 55–65, 70
Camilleri, Joseph, 120, 156, 250–51
Campaigning. *See* Nongovernmental organizations (NGOs)
Canada, 49, 86–88, 93
Capitalism, 7, 109–10, 118, 251; community resistance to, 168; historical perspective on, 127; and indigenous peoples, 39–40; and international institutions, 156–57; and knowledge, 164–66
Carcinogens, 180, 184
CFCs, 191–93
Chatterjee, Partha, 168
Cheney, Jim, 69–70
Chittagong Hill Tracts. *See* Bangladesh
Civil society, 133–34, 142, 162; and community, 127–28, 168–69
Clinton administration, 24, 27–28, 92
Coady, Tony, 16n

257

Cobb, John, Jr., 39, 79–80
Cockburn, Alexander, 166
Collingridge, David, 183
Colonialism, 47–48, 84, 111, 156, 230. See also Core-periphery
Commission on Developing Countries and Global Change, 89, 225n, 235
Community, 248–49; biospheric, 57–58, 92–94; and civil society, 127–28, 168–69; communitarianism, 81–82, 139; cultural groups, 42–43; and decision making, 41; and ecofeminism, 68; epistemic, 113, 174, 176, 186–88; and geography, 6–7, 80–85; and nonhuman elements, 53; and regionalization, 82; and resource egalitarianism, 40; social movements, 133–34; as steward of ecology, 168–70
"Concrete others," 91–94
Consciousness, 54–56, 62–63
Consumerism, 36–37, 238–40
Coordinating Body for Indigenous People's Organizations of the Amazon Basin, 114–15
Core-periphery, 83–94, 112, 249–51; decline of core, 132–33, 137; and division of labor, 127; and resources, 86–87, 90–92. See also North-South divide; Settlement
Costs: externalization of, 15; fair allocation of, 17, 97, 102; and global warming, 14, 17, 97, 232–33; of preferences, 11
Cox, Robert W., 129
Craig, Paul, 116
Critical-theoretical framework, 97, 105–7, 120–21
Cultural relativism, 43–44, 48
Cuomo, Christine, 68–69

Da Cunha, Manuela Carneiro, 40
Daly, Herman, 39, 79–80
Darwinian principles, 57–58, 65, 70
Debt, 124, 152
Decentralization, 41–42, 146–47, 163–64
Decision making: and access to information, 148; and accountability, 147–48, 171; bargaining, 12–15, 222–23; and community, 41–42; consensual, 69; integrated approach, 142–43; and participation, 147–48; and role of science, 174, 177–82; and temporal interconnectedness, 148–49

Declaration of the Developing Countries on Climate Change, 225
Deep ecology, 53, 71–74, 139, 248
Developing countries. See Core-periphery; Third World
Development. See Capitalism; Economics
Developmental state, 110–11, 154–57, 250–51. See also Dominant logic
Discourse, 193–95, 229; ecological, 157–62, 166–67
Distributive justice, 36, 230–31; and background justice, 97–98, 105; competing understandings of, 43–44; as consumerist, 36–37; and cultural relativism, 43; and indigenous peoples, 31–36; and individualism, 97, 99–102, 105; and institutions, 102–5; and nonanthropocentric view, 38, 39; and positivism, 104–5; principles of, 12; and quantification, 104–5; universal theory of, 43. See also Cultural relativism; Justice; Self-determination; Standards
Dominant logic, 66, 129–31, 136; and capitalism, 156; interpretations of, 132–33; and regime theory, 137; and world system, 132–36. See also Developmental state; Hegemony
Donner, Wendy, 140, 248
Douglas, Mary, 179–80
Dowie, Mark, 211
Dryzek, John, 146
Dumont, Louis, 110
Duvall, Raymond, 108

Earth Summit. See Rio Conference on the Environment and Sustainable Development
Eaton, Randall, 71
Eckersley, Robyn, 36
Ecocentrism, 38–40, 72, 139, 247; and value, 59–60, 62–65, 140
Ecofeminism, 53, 65–71, 248; and deep ecology, 71–74; and dualism, 66–67; ethic of care, 68–70; and interconnectedness, 66–67
Ecological Society of America, 190
Economics, 30–31, 117–18; and dislocation, 154–57; earthrights model, 236–37; and epistemic communities, 191–92; equitable commons model, 234–35, 238; fiduciary trust model, 236–37; Fordism, 118; funding, 130, 227–28; and geography, 78, 80; person as package model,

Economics (*cont.*)
237–41; realist model, 233–34; and regime change, 142–43; and state, 110–
11, 117, 151–53; and Third World exploitation, 130–31, 154–57; utilitarian
market argument, 231–33; and world
system, 127, 134
Ecosystems, 5–6, 72; differentiation of,
75–76; inherent rights of, 38–39; moral
standing of, 37–38; and net primary organic productivity, 76–77
Ehrlich, Anne, 239
Ehrlich, Paul, 239
Elites, 31–34, 84, 132
Emissions: allocation of, 17, 22, 102; per
capita, 23, 235, 240–41; stabilization
plan, 237–38; subsistence level of, 23–
25. *See also* Global warming
Environmental degradation: dimensions of,
122–24; geographical location of, 125–
26; poverty as source of, 124–25. *See
also* Impoverishment
Environmental Protection Agency (EPA),
177–78, 184
Epistemic communities, 113–14, 174–76,
186–88; and authority, 195–97; and
causal framework, 189; and commitment, 190 91; and disciplinary interests,
189–90; and domestic policy, 192–93;
and economic interests, 191–92; and
factual knowledge, 188–89; and hegemony, 193–95
Equality argument, 45–48
Ethics: cornucopian, 139; egocentric, 139;
extrication, 16–17; and geography, 94–
95; global, 223; and guidance, 64–65;
homocentric, 139; human-centered, 3–4;
and management, 53, 56, 134–35, 144–
45, 243–44; technocentric, 139, 170.
See also Deep ecology; Ecofeminism;
Land ethic; Standards
Ezrahi, Yaron, 194

Fairness, 5–6, 200; and allocation of costs,
17, 102, 231; and global warming, 13–
18; standards for, 12–13. *See also* Dis-
tributive justice
Falk, Jim, 120
Federalism, 41–42
Feminism. *See* Ecofeminism
First World: obligations of, 33, 44, 49,
87–88; settlement of, 34–35. *See also*
Capitalism; Economics

FoE. *See* Friends of the Earth
For Earth's Sake. See Commission on De-
veloping Countries and Global Change
Fox, Warwick, 72
Framework Convention on Climate
Change (FCCC), 103, 121, 225–30;
Conference of the Parties (COP), 226–
27, 234; funding, 227–28
Fraser, Nancy, 169
Friends of the Earth (FoE), 203–4, 207–8,
211, 215–19, 228
Future generations, 8, 10, 19, 149

Gadgil, Madhav, 161
Galston, William, 104
Geography (Space, Place), 150; core-
periphery relationships in, 83–88; scientific treatment of, 78–79; and self, 78;
and social movements, 166–67; and universalism, 78–80
Glasser, Robert, 116
Global commons, 144–45, 149, 214, 222–
23
Global environmental issues, 1, 245; and
developing countries, 224–25, 229–30;
joint implementation strategies, 222,
230, 240
Global Environment Facility (GEF), 227–
28
Global warming, 1–6, 9; costs of prevention, 13–14, 17, 97, 232–33; and economic activity, 13–14, 25, 27–28; and
fairness, 13–18; greenhouse warming
potential (GWP), 241; mitigation of, 25–
26; and non-carbon technology, 27–28;
and transition/extrication period, 20–29.
See also Emissions
Gong, Gerrit, 112
Goodpaster, Kenneth, 56–57
Gramsci, Antonio, 119, 162
Green movement, 40–41, 192–93, 210
Green Party, 212, 215
Greenpeace, 202–5, 210, 213–14, 217,
220
Greenpeace International Science Unit, 203
Grubb, Michael, 13n, 17n, 23n, 27n,
151n, 241n
Guha, Ramachandra, 161

Haas, Peter, 113, 186–88
Habermas, Jürgen, 106–7, 229
Hampson, Fen Osler., 226–27, 234
Hardin, Garrett, 59, 234n

Harvey, David, 167
Heartland-hinterland relationships. *See*
Core-periphery
Hecht, Susanna, 166
Hegemony, 119, 162; and counter-
hegemony, 119–20; and epistemic com-
munities, 193–95; and geography, 82–
83; and social movements, 162–63. *See
also* Dominant logic
Holism, 57–62, 248; and individualism,
58–60
Hume, David, 100, 103
Hurrell, Andrew, 101

Ideology: counterhegemonic, 119–20; received, 159; and state, 109–10, 117–19
Impoverishment: and distributive process,
124; environmental causes of, 123–26;
and food production, 125; and regime
change, 136–39, 142–43; and resources,
124–25; state role in, 126, 129–31
Independence, 108–9
India, 154–55, 160, 163–64, 167–68
Indigenous peoples, 3, 6–7, 247, 250; and
decentralist argument, 41–42; and development, 39–40; and distributive justice,
31–36; and dominant logic, 131; inherent rights claims of, 35–36; and international law, 48–49; as minority culture,
44–47, 49–50, 247–48; and modernist
values, 115–16; and net primary organic
productivity, 76–77; and periphery, 90–
91; and prior occupancy, 47; and progressive economization, 46, 49; and self-
determination, 48–50, 84–88; and settlement, 31–36; and social movements,
158, 166–67; and technology, 214;
views of environment, 114–15
Individualism, 53, 106, 249; and community, 40–41, 106; and distributive justice, 97–102, 105; and equality, 3–6;
and holism, 58–60; individual-in-community, 40–41; individual-in-culture,
42–43; and moral extensionism, 56–57;
and possessiveness, 99–101, 110; and
state, 100–101
Inherent rights, 35–36, 38–39
Inherent value, 58, 62–65, 248
Institutions: and capitalism, 155–56, 171;
constitutive importance of, 108–9; and
distributive justice, 102–5; and environmental policymaking, 116, 120; and
global social contract, 144–45, 251;

hierarchy of, 107–11; and legitimacy,
112; levels of, 103, 107–8; multilayered,
145–47; and regime theory, 137
Intergovernmental Panel on Climate
Change (IPCC), 27, 208, 226, 228–29,
237–38, 242; Special Committee, 229
International Fund for Animal Welfare
(IFAW), 214
International Monetary Fund (IMF), 113,
130, 142, 150
*International Politics of the Environment,
The* (Hurrell, Kingsbury), 101

James Bay, 64
Jasanoff, Sheila, 113, 134, 140, 201, 252
Jonas, Hans, 243
Justice: background, 96–98, 103, 105–6;
critical theory, 97, 105–7; and geogra-
phy, 79–80; internal, 96–99; interna-
tional, 96–97; theories of, 6–7, 246–49.
See also Distributive justice; Fairness

Katz, Eric, 38
Kempton, Willett, 116
Kennedy, Paul, 238–39
Kingsbury, Benedict, 101
Knight, David B., 11, 146, 248–49
Knowledge, 164–66; and authority, 195–
97; and epistemic communities, 188–89;
and state, 112–15
Kothari, Smitu, 251
Kratochwil, Friedrich, 104–5
Kuhn, Thomas, 138, 183
Kymlicka, Will, 146, 247–49

Land ethic, 53, 57–62; and animal rights,
61; and evolutionary scale, 59; and over-
population, 59–61
Latour, Bruno, 183
Law of the Sea Convention, 144–45, 226
Lean, Geoffrey, 203–4
Leggett, Jeremy, 203
Leopold, Aldo, 53, 57, 59, 71
Liberalism, 5–6, 36, 43, 45–46, 249; and
universalism, 40, 46. *See also* Individual-
ism
Localization (Microregionalization), 151

McCarthy, Thomas, 89
McCormick, John, 210
Marglin, Frederique, 165
Maximum sustainable yield, 243
Merchant, Carolyn, 110, 139

Merton, Robert, 191
Meyer, John, 119
Milieu Defensie, 218–20
Minority cultures, 44–47, 49–50, 247–48.
   See also Indigenous peoples
Montreal Protocol for the Protection of the
   Ozone Layer (1987), 14, 103, 131–33,
   188, 193, 226–27, 247
Moore, G. F., 62
Moore, Patrick, 202
Moral extensionism, 54–57
Moral standing, 7–8, 18–19, 248; and
   biotic community, 58; and conscious-
   ness, 54–56; of ecosystems, 6, 37–38;
   holism, 55–56; and human sympathy,
   58–59, 65, 70; and inherent value, 58;
   and legitimacy of actors, 11–12, 98–99,
   111–16; nonanthropocentric, 37–39; of
   nonhuman elements, 52–53
Mumford, Lewis, 194
Murphy, Alexander, 82
Myers, Norman, 33

Naess, Arne, 52, 71–72
Nairobi Declaration (1982), 136
Narain, Sunita, 23, 235, 241–42
Nation. See State
National Research Council, 190
Naturalistic approach, 18
Nature, 39; and science, 115, 202; uncertainty of, 242–43; Nature's Veto, 222–
   23
Net primary organic productivity (NPP),
   11–12, 76–78, 86
New Delhi Conference, 225
Nongovernmental organizations (NGOs),
   193, 222, 252–53; activities of, 205–6;
   and authenticity, 201; cautiousness of,
   210–11; conservation groups, 202, 205;
   dependence on science, 204–9, 213,
   219; and expertise, 199, 210; and feedback, 209; and future limits, 218–19;
   and international campaigns, 211–20;
   lay support for, 210; media image of,
   203–4; Northern, 214–17, 224–25,
   235; and power imbalance, 200–201;
   scientification of, 201–5, 209–11
Nordhaus, William, 231–33, 244
North-South divide: and geography, 87,
   89–93, 126; and NGOs, 214–17, 224–
   25, 235. See also Capitalism; Core-
   periphery; Economics

Obligations, 9, 33, 44, 87–88
Oechsli, Lauren, 38
Ophuls, William, 146
O'Riordan, Timothy, 139
Other: concrete, 88–94; generalized, 82–
   83, 93; indigenous peoples as, 91; and
   North-South split, 89–90
Ozone depletion, 188, 192–93, 195–96,
   226–27, 228

Pacifist's fallacy, 16–17
Paradigm, 183; and shifts, 116–20, 137–
   40
Parikh, Jyoti, 237–38
Parry, Martin L., 26
Participatory governance, 7–8, 254
Parti Quebecois (PQ), 86
Pateman, Carol, 169
Patriarchy, 66–67
Pearce, David, 231–33
Pearce, Fred, 203
Place. See Geography
Plumwood, Val, 53, 73–74
Polluter pays principle, 15–16, 247
Population, 4–5, 59–61, 124, 215–16,
   225
Porritt, Jonathan, 212, 215
Porter, Gareth, 205
Positivism, 104–5
Postmodernism, 200
Poverty, Population and the Planet
   (Friends of the Earth), 215–16
Precautionary principle, 223, 243
Preferences, 9–11; and state ideology, 28–
   29; and voluntaristic approach, 25
Property rights, 47–48, 100, 110–11
Public-interest groups. See Nongovernmen-
   tal organizations (NGOs)

Radical environmentalism, 52–53
Rawls, John, 79, 107
Reeve, Colin, 183
Reflexive modernization, 199–201
Regan, Tom, 53, 55–56, 59, 62, 66, 70
Regime change: conditions for, 117–20,
   138–39, 172; and dominant logic, 137;
   and economics, 142–43; and just bar-
   gains, 145; and redefinition of norms,
   137–42; and social contract, 144–45,
   251; state role in, 149–53
Regionalization, 150–51
Reilly, Charles A., 158

Resource egalitarianism, 32–37, 40; and nonindigenous peoples, 49–50
Resources: capitalist control of, 156–57; and core-periphery relationships, 86–87, 91–92; as human-centered, 36, 52; and preferences, 10–11
Responsibility, 236–37, 247, 254–55; causal, 13–15, 19–20; voluntaristic approach to, 18–20
Reus-Smit, Christian, 13, 28, 127, 129–30, 250–51
Revolution, 158, 166
Rio Conference on the Environment and Sustainable Development (Earth Summit), 1, 88, 135–36, 143, 212, 215, 220–21; *Agenda 21*, 135, 143, 221
Rio Declaration on Environment and Development (1992), 14–15, 135, 149–50, 221, 247
Risk, 179–80, 236
Rosenzweig, Cynthia, 26
Royal Society for Nature Conservation (RSNC), 206–7
Royal Society for the Protection of Birds, 216
Ruggie, John, 105

Sand County Almanac, A (Leopold), 57
Sandel, Michael, 81–82
Schaffer, Simon, 194
Schelling, Thomas C., 233–35
Science, 25–26, 252–53; and boundary-shifting, 181, 190–91; and closure, 182–84; critique of, 199–200; and geography, 78–79; and paradigm shifts, 139–40; popular view of, 175; preservation of nature for, 115, 202; realist policy, 184–86; and risk perceptions, 179–80; and social constructivism, 175, 177–82, 189; and state, 112–15; and trans-science, 178, 181; uncertainty of, 178–79, 181–82, 186–88; as universal discourse, 7, 173–74, 252. *See also* Knowledge
Science advising, 173–84; and campaigning, 209–11; and counterexperts, 199, 200; and policy, 177–84; and scientific activity, 199
Sebenius, James, 188
Secessionism, 50
Self, 53; alienation of, 72–73, 78; and geography, 95; self-in-relationship, 73–74
Self-determination, 47–50, 84–86, 88

Sessions, George, 72
Settlement, 31–36, 88; ecocentric argument against, 38–39; flaws in plans, 34–35; frontier mythology, 34, 44. *See also* Core-periphery
Shallow ecology, 52, 71–72
Shapin, Steven, 194
Sharma, B. D., 155
Shue, Henry, 96, 102, 107, 126, 144, 172, 230–31, 241, 246–47, 253
Singer, Peter, 53–56, 60
Smith, Adam, 109–10
Social movements: and biodiversity, 159–60; and community, 133–34; contradictions between, 158–59; development of, 160–61; and ecological discourse, 166–67; and geography, 166–67; and hegemony, 162–63; in India, 160, 167–68; and state change, 119–20
Societal ontology, 105–7, 250
Socio-ecological crisis, 126–36; and dominant logic, 132–33
Soto, Alvaro, 126
Sovereignty, 84–85, 98, 101, 108–9, 149–50; and change, 116–17; as weapon of state, 163
Space. *See* Geography
Speciesism, 55
Sport hunting, 61, 70–71
Stairs, Kevin, 209, 217
Standards: for fairness, 12–13; fault-based, 13, 15–16, 19–20, 247; long-range, 20–22; minimum, 20–22, 29; no-fault, 13, 15, 247; and responsibility, 13–15; short-range, 20–22; specific *vs.* general, 13; for transitions, 13, 16–17, 20–29
State, 249–51; actors within, 7–8, 98–99, 111–16; alternatives to, 163–64; and change, 116–20; and civilization, 112–13; as community, 85; and ecological revolution, 110–11; economic function of, 110–11, 117, 151–53; and geography, 84–85; and hierarchy of institutions, 107–11; historical perspective on, 126–27; Hobbesian, 30, 108; and ideology, 109–10, 117–19; and individualism, 100–101; and inequality, 14, 22–23, 76–78, 154–55, 230–31; management capability of, 134–35, 144–45; moral purpose of, 7, 28, 98, 107–16, 129, 250; normative arenas of, 109, 119–20, 153; and paradigm shifts, 141–42; peripheral, 130–31; and regime

State (*cont.*)
  change, 149–53; and resource exploitation, 111; responsibilities of, 149–50;
  and scientific knowledge, 112–15; and
  self-determination, 85–86; and social
  movements, 119–20; and world system,
  126–27. *See also* Sovereignty
Stein, Arthur, 100
Stockholm Conference (1972), 14, 135–36,
  224, 247
Sumner, L. W., 59
Sustainable development, 34, 136, 139–41,
  159, 176, 185–86, 236–37

Taylor, Charles, 106
Taylor, Pete, 209, 217
Technology, 27–28, 134, 153, 194–95,
  214; and ethics, 139, 170
Third World: and consumption patterns,
  238–40; economic exploitation of, 130–
  31, 154–55; and global agreements,
  229–30; NGO policies toward, 214–17;
  and overpopulation, 60–61; position on
  environment, 224–25; right of development in, 30–31; role of elites in, 31, 33–
  34; view of First World agenda, 35, 92–
  93
Timmerman, Peter, 253
Transition/extrication, 13, 16–17, 20–29
Turner, R. Kerry, 139

United Kingdom, 202–7, 210
United Nations, 213; *Agenda 21*, 135,
  143, 221; Cairo population conference
  (1994), 215; Conference on Environment
  and Development (UNCED), 132–33,
  135, 156; environmental funds, 133;
  Group of 77, 132–33; *Human Development Report,* 89; transfer of power to,
  135–36. *See also* Global Environmental
  Facility; Rio Conference on the Environment and Sustainable Development

Universalism, 43, 106–7, 165, 203; and
  geography, 78–80; and individualism,
  40, 45–46; and science, 7, 173–74, 252
Utilitarianism, 54

Value: and consciousness, 63–64; hierarchy of, 63, 66–68, 140; and indigenous peoples, 115–16; inherent, 58, 62–
  65, 248; and property rights, 110–11,
  119; of species, 59–60
Voluntaristic approach, 18–19, 25

Wallace, Alfred Russel, 176–77, 195
Wallace, Iain, 11, 146, 248–49
Wallerstein, Immanuel, 126–27
Walzer, Michael, 43–44, 97, 104
Warren, Karen, 53, 66, 69, 74
Wealth of Nations, The (Smith), 110
Weber, Max, 118
Weinberg, Alvin, 178–80
Wendt, Alexander, 108
Wildavsky, Aaron, 179–80
Wilderness, 39
Winner, Langdon, 194
World Bank, 113, 130, 142, 150, 171, 227
World Commission on Environment and
  Development, 185–86
World system, 126–36, 250–51; and bio-
  sphere, 129; and civil society, 127–28;
  and dominant logic, 132–36; and eco-
  nomics, 134; levels of, 128–29; and
  world polity, 127
World Trade Organization (WTO), 156,
  171
World Wide Fund for Nature (WWF),
  216–17
Worster, Donald, 202

Yearley, Steven, 134, 140, 193, 252–53
Young, Iris Marion, 104
Young, Oran, 138, 223